Cybersecurity Discourse in the United States

This book examines the role of cyber-doom rhetoric in the U.S. cybersecurity debate.

For more than two decades, fear of "cyber-doom" scenarios – i.e. cyberattacks against critical infrastructure resulting in catastrophic physical, social, and economic impacts – has been a persistent feature of the U.S. cybersecurity debate. This is despite the fact that no cyberattack has come close to realizing such impacts. This book argues that such scenarios are part of a broader rhetoric of cyber-doom within the U.S. cybersecurity debate, and takes a multidisciplinary approach that draws on research in history, sociology, communication, psychology, and political science. It identifies a number of variations of cyber-doom rhetoric, then places them into a larger historical context, assesses how realistic the fears expressed in such rhetoric are, and finally draws out the policy implications of relying on these fears to structure our response to cybersecurity challenges. The United States faces very real cybersecurity challenges that are, nonetheless, much less dramatic than what is implied in the rhetoric. This book argues that relying on cyber-doom rhetoric to frame our thinking about such threats is counterproductive, and encourages us to develop ways of thinking and speaking about cybersecurity beyond cyber-doom.

This book will be of much interest to students of cybersecurity, foreign policy, public administration, national security, and international relations in general.

Sean T. Lawson is Associate Professor of Communication at the University of Utah and Adjunct Scholar at the Modern War Institute at West Point, USA.

Routledge Studies in Conflict, Security and Technology

Series Editors:
Mark Lacy, Dan Prince, *Lancaster University*
Sean Lawson, *University of Utah*
Brandon Valeriano, *Cardiff University*

The *Routledge Studies in Conflict, Technology and Security* series aims to publish challenging studies that map the terrain of technology and security from a range of disciplinary perspectives, offering critical perspectives on the issues that concern publics, business, and policymakers in a time of rapid and disruptive technological change.

Terrorism Online
Politics, law, technology
Edited by Lee Jarvis, Stuart Macdonald and Thomas M. Chen

Cyber Warfare
A multidisciplinary analysis
Edited by James A. Green

The Politics of Humanitarian Technology
Good intentions, unintended consequences and insecurity
Katja Lindskov Jacobsen

International Conflict and Cyberspace Superiority
Theory and practice
William D. Bryant

Conflict in Cyber Space
Theoretical, strategic and legal perspectives
Edited by Karsten Friis and Jens Ringsmose

US National Cybersecurity
International Politics, Concepts and Organization
Edited by Damien Van Puyvelde and Aaron F. Brantly

Cybersecurity Discourse in the United States
Cyber-Doom Rhetoric and Beyond
Sean T. Lawson

For more information about this series, please visit: www.routledge.com/Routledge-Studies-in-Conflict-Security-and-Technology/book-series/CST

Cybersecurity Discourse in the United States

Cyber-Doom Rhetoric and Beyond

Sean T. Lawson

Routledge
Taylor & Francis Group

LONDON AND NEW YORK

First published 2020
by Routledge
2 Park Square, Milton Park, Abingdon, Oxon OX14 4RN

and by Routledge
605 Third Avenue, New York, NY 10017

First issued in paperback 2021

Routledge is an imprint of the Taylor & Francis Group, an informa business

Publisher's Note
The publisher has gone to great lengths to ensure the quality of this reprint
but points out that some imperfections in the original copies may be
apparent.

British Library Cataloguing-in-Publication Data
A catalogue record for this book is available from the British Library

Library of Congress Cataloging-in-Publication Data
A catalog record for this book has been requested

ISBN 13: 978-1-03-208276-9 (pbk)
ISBN 13: 978-1-138-20182-8 (hbk)

Typeset in Times New Roman
by Apex CoVantage LLC

For Everett, who reminds me what is most important.

Contents

Acknowledgments   viii

1   The cyber sky is falling! An introduction to cyber-doom   1

2   From *WarGames* to cyber Pearl Harbor: motivating
    cybersecurity with cyber-doom rhetoric   36

3   From wire devils to cyber squirrels: cyber-doom rhetoric
    as fear of technology-out-of-control   77

4   Panic, paralysis, and social collapse: the exaggerated fears
    of cyber-doom   100

5   When fear fails: the dangers of cyber-doom rhetoric   127

6   Cold War 2.0 and the emergence of cyber-enabled political
    warfare   161

7   After action report and lessons learned   185

    Index   205

Acknowledgments

This book would not have been possible without the advice and support of family, friends, and colleagues. First and foremost, I must thank my wife, Cynthia, for all of her support, including help working out my thoughts, editing, and general encouragement to keep going at those moments when I felt like giving up. I would also like to thank my friends and colleagues at the University of Utah and elsewhere who have been instrumental to the completion of this book. This includes Marouf Hasian, Robert Gehl, and Michael Middleton at the University of Utah. It also includes Brandon Valeriano of the Marine Corps University. Each of them provided invaluable advice and encouragement along the way. This book also benefited from feedback received by presenting portions of it to cybersecurity students at Northeastern University and researchers at the Center for Global Security Research at Lawrence Livermore National Laboratory. Thank you to Ryan Maness and Jackie Kerr for those respective invitations.

Finally, I would like to thank the publishers who have granted permission for me to build upon some of my previous work published with them. Portions of Chapter 1 draw from Sean Lawson (2013), "Motivating Cybersecurity: Assessing the Status of Critical Infrastructure as an Object of Cyber Threats," in Laing, C., Badii, A. and Vickers, P. (eds) *Securing Critical Infrastructures and Critical Control Systems: Approaches for Threat Protection*, Hershey, PA: IGI Global, pp. 168–89. Portions of Chapters 3 and 4 draw from Sean Lawson (2013), "Beyond Cyber-Doom: Assessing the Limits of Hypothetical Scenarios in the Framing of Cyber-Threats," *Journal of Information Technology & Politics*, 10, 1: 86–103; Sean Lawson (2011), "Articulation, Antagonism, and Intercalation in Western Military Imaginaries," *Security Dialogue*, 42, 1: 39–56; and Sean Lawson (2014) *Nonlinear Science and Warfare: Chaos, Complexity, and the U.S. Military in the Information Age*, London: Routledge. Portions of Chapter 5 draw from Sean Lawson (2012), "Putting the 'War' in Cyberwar: Metaphor, Analogy, and Cybersecurity Discourse in the United States," *First Monday*, 17, 7.

1   The cyber sky is falling! An introduction to cyber-doom

Cyberspace, November 2012

In November 2012, viewers around the world were stunned when international news media reported what one network called "a cyberterrorist assault" on a major Western intelligence agency. Hackers broke into the agency's computer systems, manipulated its heating, ventilation, and air conditioning (HVAC) system to cause a natural gas leak, and then used that leak to blow up a portion of the building. In all, six were killed and many more were injured.

But this dramatic incident was actually the latest salvo in a cyber conflict that had already been raging for weeks. A disgruntled former intelligence operative had managed to steal a trove of highly classified information from the agency, in particular the names of every undercover NATO intelligence agent serving in terrorist organizations around the world. This insider-turned-cyberterrorist vowed to release the names of five agents per week on YouTube, a promise on which he ultimately made good, leading to the death of at least three agents, one of which was videotaped and posted online.

In the ensuing response to what many called the "greatest internal security breach ever," government intelligence agents always seemed one step behind. Indeed, for a time, the government was unable to solve the attribution problem as the cyberattacker used an anonymity network to route his computer traffic all over the globe, making identification and geo-location of the attacker almost impossible. Although the government was ramping up a global manhunt, it did not at first even know who it was hunting, which led to increasing criticism and public embarrassment. What's more, while the government seemed to chase shadows around the globe, the hacker continued to break into government computer systems with impunity, to carry out the "kinetic" cyberattack mentioned earlier, and even to carry out other cyber-kinetic attacks, including one that derailed a subway train.

In the end, the government was able to identify the former insider and put a stop to his campaign of cyberterror. But the cost was high. What started as a conflict in cyberspace eventually ended in physical confrontation between the government and cyberterrorist. The perpetrator was eventually killed in this final showdown, but so too were a number of government agents. Perhaps most importantly, however, one cyberterrorist had, for a time, brought a government and a nation to a virtual standstill. He had not only called into question the government's ability to secure sensitive information from massive leaks, but also its fundamental ability to provide for the safety of its people. Though this particular cyberterrorist attack ultimately failed, it nonetheless demonstrated the potentially catastrophic consequences of such tactics and focused public attention on the perils of poor cybersecurity.

Overview

The tale of cyberattack that opens this book will likely seem familiar to many readers. That is because it is actually the plot of *Skyfall*, the 2012 blockbuster installment in the James Bond series of movies. However, this fictional portrayal of cyberattack is also likely familiar because it is a powerful reflection of real and growing concerns about cybersecurity in many Western nations, the United States in particular. News media and policy makers have paid increased attention to these concerns in recent years and for good reason. It seems that not a week goes by without reports of another major data breach against a large company or even the U.S. government. Of course, Russian cyber operations to interfere in the 2016 U.S. presidential elections served to focus attention on issues of cybersecurity like never before.

But to close observers of the public discourse about cybersecurity in the United States, *Skyfall* will likely be familiar for another reason. It is reflective of the tendency for fact and fiction to blur, and, in particular, the prominence of what some have called "cyber-doom scenarios" in public discourse about cybersecurity (Dunn Cavelty, 2008: 2). Such scenarios, this book will argue, are an important element of a wider rhetoric of cyber-doom that is a persistent feature of U.S. cybersecurity discourse and Information-Age security imaginary.

Cyber-doom scenarios involve hypothetical tales of cyberattacks, often against critical infrastructures like power, water, or transportation systems, resulting in mass destruction or disruption. The imagined impacts of such attacks sometimes include serious economic loss, property damage, or loss of life. Many scenarios contemplate secondary or tertiary effects that include total economic, social, or civilizational collapse. News media, pundits, policy makers, military officers, and even some religious leaders regularly deploy such scenarios when talking about cybersecurity (Rhodes, 2011; Lawson and Middleton, 2019). But as this book will argue, such scenarios are just one of several tactics of a broader cyber-doom rhetoric that has the effect of misdiagnosing and, in the process, inflating the impacts of the very real cybersecurity challenges that we face.

Cyber-doom scenarios and rhetoric are not limited to Hollywood or science fiction. Rather, they are common in serious public policy discussions about cybersecurity. Just one month before the release of *Skyfall*, Secretary of Defense Leon Panetta gave a speech warning of the real potential of a "cyber Pearl Harbor" and offered a cyber-doom scenario that was even more fantastical than the one portrayed in *Skyfall*. He asked his audience of Business Executives for National Security to imagine the worst:

> An aggressor nation or extremist group could gain control of critical switches and derail passenger trains, or trains loaded with lethal chemicals. They could contaminate the water supply in major cities, or shut down the power grid across large parts of the country.
>
> The most destructive scenarios involve cyber actors launching several attacks on our critical infrastructure at once, in combination with a physical attack on our country. Attackers could also seek to disable or degrade critical military systems and communications networks.
>
> The collective result of these kinds of attacks could be "cyber Pearl Harbor": an attack that would cause physical destruction and loss of life, paralyze and shock the nation, and create a profound new sense of vulnerability.
>
> (Panetta, 2012)

Panetta called his scenario "shock therapy." Adam Stone of *Fifth Domain* described it as an attempt "to shake up ordinary citizens, to awaken them to the seriousness of the situation." Similarly, Stone reports that Edward Wittenstein of Yale University's Johnson Center for the Study of American Diplomacy commented, "It was a clarion call, evoking the specter of a horrific national disaster as a way to galvanize public support. It also communicated that urgency to members of Congress, whose support was needed for solidifying and resourcing Cyber Command." Secretary Panetta's speech, Stone reports, "would help to shape the nation's cyber strategy in the short term, and the ramifications of that warning continue to be reflected in Cyber Command's evolving mission today" (Stone, 2019).

Though Secretary Panetta's comments garnered a great deal of attention, he was not the first or last to have used the Pearl Harbor metaphor to warn of impending cyber-doom. Cyber Pearl Harbor in particular has been a feature of the U.S. public policy debate about cybersecurity for over twenty-five years and has, from the beginning, most often referred to doom-like scenarios resulting from cyberattacks against critical infrastructure (Lawson and Middleton, 2019). In the metaphor's first known usage (Stevens, 2016: 131), for example, computer security entrepreneur Winn Schwartau described the impacts of such an event as "truly crippling," "devastating," and "inflicting massive damage" on a scale that would undermine "the continuation of well-ordered society . . . [to] function as we know it" (Schwartau, 1991b).

News and entertainment media has played an important role over the years in helping to keep cyber-doom front and center in the minds of the public and policymakers alike. In 1999, for example, *New York Times* reporter John Markoff asked readers to imagine "simultaneous computer network attacks against banking, transportation, commerce and utility targets – as well as against the military – conjures up the fear of an electronic Pearl Harbor in which the nation is paralyzed without a single bullet ever being fired" (Markoff, 1999). As early as 1983, a

Hollywood movie, *WarGames*, was central to raising awareness and promoting legislative action on cybersecurity (Schulte, 2013; Kaplan, 2016).

More recently, *Skyfall* itself became fodder in the public debate about cybersecurity, with some arguing that the cyberattack scenarios it portrayed were a real possibility (Waugh, 2012; Epstein, 2012; Pierce, 2012; Yakabuski, 2012), while others claimed that they were unrealistic and contributed to a counterproductive climate of fear about all things "cyber" (Nexon, 2012; Curran, 2012; Dickinson, 2012; Hubert, 2012). There is at least some reason to take those latter concerns seriously. Indeed, for at least one otherwise serious observer, the image of fact and fiction seemed to have completely reversed.

Asked by *CNN* whether various of the cyberattack plot elements of the film were realistic, one cybersecurity expert said that while some of the details were off the mark, in general, most of the scenarios were plausible, including some of the most dramatic, like causing explosions and derailing trains. In real life, however, though there have been a few examples of physical damage caused by cyberattack, we have yet to see anything that approaches the kinds of scenarios depicted in *Skyfall*, and certainly nothing like what Secretary Panetta imagined. Conversely, when asked about *Skyfall*'s opening sequence, in which a hacker possesses a huge trove of sensitive, stolen intelligence information that he threatens to dump on the Internet for all to see, *CNN*'s cybersecurity expert found such a scenario unrealistic. Based on its interview, *CNN* reported that "it's unlikely that [the hacker] would be able to have all of this information in one place" and that "it would be virtually impossible for someone to have access to all of that information as 'everything is regulated on the government side'" (Pierce, 2012).

But of all the cyber scenarios in the movie, this is the one that was actually the most plausible because it had, at that point, already happened. And it was about to happen again. Two years prior, through the spring and fall of 2010, the U.S. government sat helpless as Julian Assange and WikiLeaks released top secret video from Iraq that seemed to show U.S. pilots carrying out airstrikes against journalists and ambulance drivers, sparking outrage and claims that the U.S. had committed war crimes. Then, in the fall, the group began releasing a trove of roughly 250,000 U.S. diplomatic cables. We later learned that just one low level insider, an Army private serving as an intelligence analyst at a forward operating base in Iraq, had been able to access all of this information from a work terminal, steal it, and then dump it to the Web for all to see, a scenario eerily similar to the one from *Skyfall* that *CNN*'s cybersecurity expert found implausible (Greenberg, 2012).

But the worst was yet to come, and it still did not involve the kinds of cyberdoom scenarios found in *Skyfall* or Secretary Panetta's speech. Just six months after *Skyfall*'s release, the world was shocked once again, this time by a series of revelations about National Security Agency (NSA) surveillance activities. These revelations came from yet another breach of secret intelligence information from another insider. Former NSA contractor Edward Snowden had managed to steal tens of thousands of secret documents and leak them to journalists who for years afterwards published regular reports embarrassing to the agency and which sparked outrage around the world (Greenwald, 2014).

Such threats were not just coming from insiders, however. In summer 2015, the world learned that the computer systems of the U.S. government's Office of Personnel Management (OPM) had been breached. Chinese cyber spies, the government claimed, had stolen information about millions of government employees, including many with top secret security clearances, raising fears about further exploitation of these individuals and further loss of secret information (Sanger, 2015a, 2015b; Davis, 2015). Then, in 2016, we began to learn of a Russian operation to interfere in the U.S. presidential election that involved the use of cyber intrusions to steal confidential campaign information that was then leaked to WikiLeaks and further promoted by state propaganda outlets and armies of social media bots and trolls who gamed search and social media algorithms (Office of the Director of National Intelligence, 2017; U.S. Senate Select Committee on Intelligence, 2018; Mueller III, 2019). In short, while some like *CNN*'s cybersecurity expert worried about the possibilities of fictional cyber-doom scenarios, they discounted the very real and still quite serious scenarios that had already occurred and were about to occur.

Despite this seemingly apparent disconnect between rhetoric and reality, cyber-doom rhetoric has remained a common feature of U.S. cybersecurity discourse. Some invoke an as-yet hypothetical "cyber Pearl Harbor" or "cyber 9/11" to argue for taking cybersecurity more seriously. For example, in a February 2015 speech on the topic of cybersecurity, though President Barack Obama focused primarily on cyber crime and data breaches, he nonetheless urged listeners "to imagine what a set of systematic cyberattacks might do" by pointing to "real-life examples" of "an air traffic control system going down and disrupting flights, or blackouts that plunge cities into darkness" (Obama, 2015). Similarly, in August 2015, Senator Susan Collins (R-ME) urged the passage of the Cybersecurity Information Sharing Act of 2015 by warning, "We are at September 10th levels in terms of cyber preparedness." Legislation would be required, she argued, "to reduce the likelihood of a cyber 9/11" (Collins, 2015). The legislation later passed (Pagliery, 2015).

Others invoke Pearl Harbor and 9/11 to inflate the impacts of the cyberattacks that have already occurred. In November 2014, we learned that Sony Pictures Entertainment had been hacked and a large amount of sensitive, internal documents, including emails, contracts, and more, were dumped on the Internet. The U.S. government blamed the hack on the government of North Korea (Sanger and Perloth, 2014). Though this cyberattack did not result in physical damage, loss of life, or even serious damage to the American economy, NSA Director Admiral Michael Rogers claimed that the incident constituted a "cyber Pearl Harbor" of the kind Secretary Panetta had warned about back in 2012. One government technology trade publication reported Admiral Rogers as defining a "cyber Pearl Harbor" as "an action directed against infrastructure within the United States that leads to significant impact – whether that's economic, whether that's in our ability to execute our day-to-day functions as a society, as a nation." The report continued, "Movie studios fit into the U.S. government's broad definition of critical infrastructure" (Lyngaas, 2015). Similarly, in summer 2015, some called the hack of OPM and theft of personal information about millions of government employees a "cyber 9/11" (Weisman, 2015), with one lawmaker even claiming that the hack was actually a more serious threat than 9/11 (Carman, 2015).

These uses of cyber-doom rhetoric came at a time when the U.S. intelligence community was beginning to realize that cyber-doom scenarios were unlikely, were not an accurate reflection of actual threats to cybersecurity, and may even be a dangerous distraction (Marks, 2019). In February and again in September 2015, Director of National Intelligence James Clapper told Congress that "the likelihood of a catastrophic attack" or "a 'Cyber Armageddon' scenario that debilitates the entire U.S. infrastructure" is "remote at this time" (Clapper, 2015a, b). Such fictions are not, as it turns out, "the real effects of a cyber war" as some have claimed (Patterson, 2010). Instead, as DNI Clapper noted, the current cyber threat is "an ongoing series of low-to-moderate level cyberattacks from a variety of sources over time, which will impose cumulative costs on US economic competitiveness and national security" (Clapper, 2015b: 2). He warned about Russian use of propaganda, of cyberattacks against the integrity of information, and foreign intelligence services use of corporate and government stores of personal information to target American citizens (Clapper, 2015b). At the same time, a group of experts writing for NATO echoed DNI Clapper's assessment, using an analysis of Russian cyber operations in Ukraine to warn that the nature of cyber conflict was turning out to be very different than we had long assumed (Geers, 2015). The next section will illustrate that this shift in official thinking was a long time in the making. Beginning in 2008, official policy documents and statements by officials began to recognize that theft of information, including private intellectual property and government secrets, and not catastrophic attacks against critical infrastructure, is the primary cyber threat (Lawson, 2013a).

Russian cyber operations against the 2016 U.S. presidential election track very closely with these assessments. As a result, a number of critics have argued that obsession with cyber-doom scenarios was at least partly to blame for the United States failing to identify and react quickly enough to the Russian cyber threat. For example, in November 2016, James Lewis of the Center for Strategic and International Studies said, "We tend to over-militarize everything and spend our time looking for a cyber 9/11, and Russia completely went around us on it. Our doctrine is very much about protecting critical infrastructure, and their doctrine is about information warfare" (McMillan and Valentino-DeVries, 2016; For similar arguments from Lewis, see Lewis, 2017; Lewis, 2018). Many others have made a similar argument since 2016 (Pollard and Devost, 2016; Nye Jr, 2016; Lipton et al., 2016; Orcutt, 2017; Pollock, 2017; Von Drehle, 2017; Rid and Buchanan, 2018; Pollard et al., 2018; Wolfe, 2018; Weinstein, 2018). Experts continued to downplay the threat of cyber-doom into 2019. For example, in January of that year, the *Bulletin of the Atomic Scientists* addressed the issue of cyber threats in their updated doomsday clock statement, echoing DNI Clapper's assessment from 2015, "Rather than a cyber Armageddon that causes financial meltdown or nationwide electrical blackouts, this [information warfare] is the more insidious use of cyber tools to target and exploit human insecurities and vulnerabilities, eroding the trust and cohesion on which civilized societies rely" (Mecklin, 2019). Though the statement did acknowledge the serious threat that cyberattacks on critical infrastructure might yet pose, Herb Lin, one of the *Bulletin's* Science and Security Board members, later said that such attacks are "very hard to do, if not impossible" (Porup, 2019). Similarly, after an April 2019 report from the U.S. Department of Energy about a so-called cyber event that caused some minor disruption to an electrical utility, former NSA analyst and industrial control system cybersecurity expert Robert Lee told reporters that though cyber threats to critical infrastructure are a real concern, doomsday-like cyber Pearl Harbor or cyber 9/11 scenarios are not (Motherboard, 2019). Harvard Professor Joseph S. Nye agreed, arguing that "deterring major states from acts like destroying the electricity grid may be easier than deterring actions that do not rise to that level. Indeed, the threat of a 'cyber Pearl Harbor' has been exaggerated" (Nye Jr, 2019).

Nonetheless, some continue to use cyber-doom rhetoric to call for greater action to protect critical infrastructure from cyberattack. We see this, for example, in an April 2019 op-ed written by Panetta (Panetta and Talent, 2019) who infamously warned in 2012 of an impending, catastrophic "cyber Pearl Harbor" (Panetta, 2012). Others have called Russian election hacking the fulfillment of decades of cyber-doom predictions. For these observers, 2016 was the fulfillment of Panetta's cyber Pearl Harbor (Chang and Osborne, 2017; Graham, 2017; Carr, 2017; Spring, 2017; Hertling and McKew, 2018). In these cases, of course, emphasis is placed on the element of surprise, not necessarily the physical destruction – of which there was none in the Russia case – typical of cyber-doom scenarios.

Why, after more than two decades of failed cyber-doom predictions and growing realizations that we have perhaps misdiagnosed the real cyber threats that we face, does cyber-doom rhetoric persist in U.S. public policy discourse? What are the characteristics of cyber-doom rhetoric? Why is it so attractive as a tool of persuasion? What are the potential impacts of relying on such rhetoric to frame the debate about cybersecurity? How might we find alternative ways of thinking and speaking about cybersecurity that would allow us to move beyond the rhetoric of cyber-doom?

These are the questions that this book seeks to answer. It examines the use of cyber-doom rhetoric as a way of elucidating the often blurry line between fact and fiction in U.S. public discourse about cybersecurity. As mentioned earlier, a number of scholars and other observers have noted the persistence of cyber-doom rhetoric in public discourse about cybersecurity. But none have focused specifically on such rhetoric, including its history, variations, purpose, and effects. Similarly, Winn Schwartau, a cyber-doom entrepreneur who claims to have coined the phrase "electronic Pearl Harbor" (Schwartau, n.d.), suggested that confusion between fiction and fact "doesn't matter" because "they're the same after all" (Schwartau, 1991a). But even those who are not complacent about threats to cybersecurity see a potential danger in confusing the two. For example, Myrian Dunn Cavelty warns against complacency in the face of cyber threats, but also that "hysterical doomsday scenarios" are not only unlikely, but a potentially counterproductive way of framing those threats (Dunn Cavelty, 2008: 144). Taking cybersecurity threats seriously requires maintaining a commitment to sorting out the differences between fact and fiction and avoiding the temptation to act based on fear alone.

This book provides a critical appraisal of cyber-doom rhetoric that addresses questions of why, how, and with what potential impacts such rhetoric is used in U.S. cybersecurity discourse. In doing so, it takes a multidisciplinary approach that draws from research in history, literature, sociology, communication, psychology, and political science. It identifies a number of variations of cyber-doom rhetoric, places this rhetoric into a larger historical context, assesses how realistic the fears expressed in such rhetoric are, and draws out the policy implications of relying on these fears to structure our thinking about and response to cybersecurity challenges.

It argues that cyber-doom rhetoric in a number of variations is used primarily as a tool for motivating a response to a wide array of very real and serious cybersecurity challenges that are, nonetheless, much more mundane than what is implied in such rhetoric. Additionally, it argues that cyber-doom rhetoric is not entirely unique but is, rather, the latest manifestation of longstanding fears of technology-out-of-control in Western societies, as well as a tradition of blurring distinctions between fact and fiction when thinking about and responding to new technologies. From this perspective, the book argues that the specific fears expressed in cyber-doom rhetoric are not only unrealistic, but that relying on them to frame our thinking about and motivate a response to the threats we do face is counterproductive. It therefore encourages us to develop a way forward for thinking and speaking about cybersecurity beyond the use of cyber-doom rhetoric.

The remainder of this chapter will provide a brief history of cyber threat perceptions in the United States and, in particular, a recent shift towards "informational" cyber threats as the primary concern for policymakers. Next, it will explain the approach to studying cyber-doom rhetoric taken in this book and the importance of studying this rhetoric as an element of public discourse about cybersecurity. Finally, the chapter ends by providing an outline and description of the book's remaining chapters.

The evolution of U.S. cyber threat perceptions

Cybersecurity is not a new concern. Rather, it has grown over the last three decades as modern societies have become increasingly dependent upon networked information and communication systems. During that time, however, perceptions of precisely what it is that is threatened, by whom, and with what potential impacts, in and through cyberspace, have shifted. As dominant perceptions of cyber threats have shifted over time, so have claims about the primary subjects (e.g. foreign spies, criminals, terrorists, insiders), objects (e.g. business data, state secrets, critical infrastructure), and impacts (e.g. monetary loss, diminished competitiveness, catastrophe) of those threats.

In the United States, cybersecurity concerns date to the 1980s and were focused on what Dunn Cavelty has called "the hostile intelligence threat" (Dunn Cavelty,

2008: 41)—i.e. the potential for foreign espionage conducted via exploitation of networked information and communication systems. But then, in the mid-1990s, cyber threats and critical infrastructure were linked as a national security concern. In the United States, this was primarily a result of the 1995 Oklahoma City bombing (Dunn Cavelty, 2008: 91). One response to the attack was the formation of the President's Commission on Critical Infrastructure Protection (PCCIP). The 1997 report of a PCCIP study that assessed the vulnerabilities in and threats to U.S. critical infrastructures clearly linked cyber threats and critical infrastructure protection (CIP) and has remained one of the "most influential of all studies" on the subject (Dunn Cavelty, 2008: 117). Thus, in the late 1990s, the dominant U.S. perception of cyber threats underwent a transformation. As prospective cyber threats to critical infrastructure became predominant, "the focus on the foreign intelligence threat further decreased" (Dunn Cavelty, 2008: 115). As the object of the perceived threat shifted from state secrets to civilian critical infrastructure, so did the supposed subject of the threat. Where the "hostile intelligence threat" had perceived foreign intelligence agencies as the primary threat subjects, from the mid-1990s through the presidency of George W. Bush, terrorist groups were identified as most likely to engage in cyberattacks against critical infrastructure (Dunn Cavelty, 2008: 103; Conway, 2008).

But it was not just the subjects and objects of the cyber threat that underwent a transformation in the mid-1990s. Fear about the potential impacts of cyberattacks increased exponentially as critical infrastructure emerged as the primary object of such attacks. The linking of cyber threats and critical infrastructure meant "the magnitude of the threat was expanded considerably" as it was "linked to the possible destruction of the whole of society. As a consequence, cyber threats are treated as being equally dangerous as nuclear weapons" (Dunn Cavelty, 2008: 98). It is in this context that we see the emergence of cyber-doom scenarios in news media, among experts, and in policy documents (Conway, 2008: 113; Dunn Cavelty, 2008: 2; Marsh, 1997: 17–18). We will return to these early examples of cyber-doom rhetoric in Chapter 2.

Though there were a number of shifts in assessments of the primary subject of the cyber threat during the administration of President George W. Bush, critical infrastructure retained its place as the primary object of concern. In the opening months of the Bush administration, attention shifted briefly from non-state to state-level cyber threats, back to non-state actors in the wake of the terrorist attacks of September 11, 2001, and then to states once again in the run-up to war with Iraq in 2003 (Weimann, 2005: 133–4; Bendrath, 2001; Bendrath, 2003; Dunn Cavelty, 2008). While the wars in Afghanistan and Iraq led to a brief decline in U.S. public policy discussion of cyber threats, a number of high-profile cyberattacks coinciding with the shift in U.S. presidential leadership from George W. Bush to Barack Obama led to renewed interest. These included two large-scale cyberattacks, one against the country of Estonia in 2007 (Evron, 2008; Blank, 2008) and the other against the country of Georgia in 2008, both of which were widely believed to have been the work of state-sanctioned Russian hackers (Korns and Kastenberg, 2008; Nichol, 2008; Bumgarner and Borg, 2009). Indeed, one of President Obama's first major national security initiatives was to order a comprehensive review of U.S. cybersecurity policy (The White House, 2009a).

Though there were shifts in perceived cyber threat subjects during the period between 1997 and 2008, the prevailing cyber threat perception was consistent in its focus on critical infrastructure as the object of prospective cyberattacks (Dunn Cavelty, 2008: 90). But with the election of President Barack Obama in November 2008, the cyber threat perception began to shift once again. As a result, critical infrastructure slipped from its position as the primary object of perceived cyber threats. The first indication of this shift appeared in December 2008 when the Center for Strategic and International Studies (CSIS) released a report titled, *Securing Cyberspace for the 44th Presidency* (Langevin et al., 2008). The report was the result of a yearlong effort by the CSIS Commission on Cyber Security for the 44th Presidency, which was co-chaired by two Congressional representatives, James Langevin (D-RI) and Michael McCaul (R-TX); the Corporate Vice President for Trustworthy Computing at Microsoft, Scott Charney; and a retired United States Air Force officer, Lt. General Harry Raduege. James Lewis, the head of the CSIS Technology and Policy Program and a respected expert on cybersecurity policy, directed the project. After "examin[ing] existing plans and strategies" for cybersecurity, the Commission "assess[ed] what a new administration should continue, what it should change" (Langevin et al., 2008: Preface). The report clearly identified a need for change in perceptions of the primary object of cyber threats:

> In 1998 [sic], a presidential commission [PCCIP] reported that protecting cyberspace would become crucial for national security. In effect, this advice was not so much ignored as misinterpreted—we expected damage from cyber attacks to be physical (opened floodgates, crashing airplanes) when it was actually informational.
>
> (Langevin et al., 2008: 12)

But what is "informational" damage? The report focused primarily on theft of intellectual property and government secrets. It argued that "the immediate risk lies with the economy" and that the U.S. had already suffered from the theft of billions of dollars' worth of intellectual property and government secrets, including crucial data related to military technologies. Ultimately, the report warned that "America's power, status, and security in the world depend in good measure upon its economic strength; our lack of cyber security is steadily eroding this advantage" (Langevin et al., 2008: 13). Although the report acknowledged that "exploiting vulnerabilities in cyber infrastructure will be part of any future conflict," it was quick to provide the caveat that "depriving Americans of electricity, communications, and financial services may not be enough to provide the margin of victory in a conflict, but it could damage our ability to respond and our will to resist" (Langevin et al., 2008: 13). Informational threats were presented as having already been realized and as having had identifiable impacts. Infrastructural threats, while not discounted as a possibility, were framed as existing only in the future and exhibiting uncertain potential impacts. In short, the report stepped back from the cyber-doom fears of the 1990s and early 2000s.

Two years later, after being confirmed as the first commander of the newly formed U.S. Cyber Command, Gen. Keith Alexander acknowledged that the CSIS report "served as a key thread of continuity across two administrations and really set the foundation for crafting this administration's strategy for cyber and security" (Alexander, 2010). While Gen. Alexander is correct that the CSIS report was an important influence on the new administration's vision of cybersecurity, it did not represent a "thread of continuity," but rather, the beginning of an important shift in what had otherwise been a stable perception of the primary object of cyber threats. Official Obama administration statements of cybersecurity policy and strategy, both civilian and military, echoed the CSIS framing of the primary objects and impacts of cyber threats and focused primarily on the negative economic impacts of stolen intellectual property and government secrets.

The impact of the CSIS report was seen only five months after its publication, in May 2009, when the Obama administration released its *Cyberspace Policy Review*. While the overall goal of the review was to set the stage for formulating and implementing policies and strategies for "assuring a trusted and resilient information and communications infrastructure," much of the document's "case for action" focused on the threat to intellectual property and economic competitiveness. The CSIS report was referenced directly on the first page of the White House document, which was prefaced by a statement that framed the cybersecurity threat in exactly those terms outlined by the CSIS report:

> Our digital infrastructure has already suffered intrusions that have allowed criminals to steal hundreds of millions of dollars and nation-states and other entities to steal intellectual property and sensitive military information. Other intrusions threaten to damage portions of our critical infrastructure. These and other risks have the potential to undermine the Nation's confidence in the information systems that underlie our economic and national security interests.
>
> (The White House, 2009a: i)

As in the CSIS report, threats to intellectual property and military secrets leading to monetary loss were identified as having already occurred, while threats to critical infrastructure had yet to be realized but were still of concern. Variations of the phrase "economic and national security," which can be read as putting economy first and security second or as conflating the two, were used consistently throughout the *Cyberspace Policy Review*. As evidence to support its "case for action," the document cited two economic-related threats – "exploiting global financial services" and "systemic loss of U.S. economic value" – in addition to "failure of critical infrastructures" (The White House, 2009a: 2). President Obama's speech introducing the *Cyberspace Policy Review* reiterated this framing. He led the speech by highlighting the costs of stolen credit card information, stolen intellectual property, and manipulations of the financial system emanating from cyberspace to emphasize that "America's economic prosperity in the 21st century will depend on cyber security." Only then did he turn to the potential for cyber threats to civilian critical infrastructures and military information and communication networks (The White House, 2009b).

This framing served as the foundation of the United States' vision for global "Internet freedom" and norms of international behavior in cyberspace during the Obama years. In her January 2010 speech on "Internet freedom," Secretary of State Hillary Clinton made special note of Google's accusations that it had been the victim of a Chinese government cyberattack. While she noted the increasing use of Internet filtering and censorship worldwide as a means of repressing political dissent and violating religious freedoms, she used the Google case to warn, "Our ability to bank online, use electronic commerce, and safeguard billions of dollars in intellectual property are all at stake if we cannot rely on the security of our information networks." Addressing these threats, she said, "demand[s] a coordinated response by all governments, the private sector, and the international community" (Clinton, 2010). Critical infrastructure went unmentioned in her speech.

It is not surprising, therefore, that "respect for property," including "respect for intellectual property rights, including patents, trade secrets, trademarks, and copyrights," was listed as a key principle upon which to build international norms of cyberspace behavior in the Obama administration's May 2011 *International Strategy for Cyberspace* (The White House, 2011: 10). While protecting critical infrastructure by "reduc[ing] intrusions into and disruptions of U.S. networks" was identified as a policy priority (The White House, 2011: 17), economic priorities, including "protect[ing] intellectual property, including commercial trade secrets, from theft" were identified first and foremost. This prioritization of economy and property is rooted in the belief that

> Cyberspace can be used to steal an unprecedented volume of information from businesses, universities, and government agencies; such stolen information and technology can equal billions of dollars of lost value. [. . .] Results can range from unfair competition to the bankrupting of entire firms, and the national impact may be orders of magnitude larger. The persistent theft of intellectual property, whether by criminals, foreign firms, or state actors working on their behalf, can erode competitiveness in the global economy, and businesses' opportunities to innovate.
>
> (The White House, 2011: 17–18)

Finally, the economic framing of the objects and impacts of cyber threats was at the heart of a number of pieces of cybersecurity legislation proposed in the U.S. Congress during the Obama administration. Senators Sheldon Whitehouse (D-RI) and Jon Kyl (R-AZ) explained the need for their proposed Cyber Security Public Awareness Act of 2011 by claiming that cyberattacks have not only resulted in stolen intellectual property and government secrets, but also the "loss of countless American jobs" (Committee on Homeland Security and Governmental Affairs 2011a, 2011b). In an accompanying op-ed piece, Senator Whitehouse argued that while cyberattacks have the "potential" to "sabotage our critical infrastructure," cyberattacks have already put the United States "on the losing end of what could be the largest illicit transfer of wealth in world history" (Whitehouse, 2011). Similarly, the Cyber Security and American Cyber Competitiveness Act of 2011, which was introduced by Senate Majority Leader Harry Reid (D-NV), echoed Senators Whitehouse and Kyl's concern, claiming, "Businesses in the United States are bearing enormous losses as a result of criminal cyberattacks, depriving businesses of hard-earned profits that could be reinvested in further job-producing innovation" (Committee on Homeland Security and Governmental Affairs, 2011a).

Perhaps most surprising is that this focus on short-term monetary and long-term economic impacts of stolen intellectual property and government data was just as pronounced in statements by U.S. military leaders, both civilian and uniformed, as well as in official Department of Defense cybersecurity policy documents. In fact, in many cases they were even more explicit than civilian policymakers in stating that cyber threats to private intellectual property and sensitive government data have taken precedence over cyber threats to critical infrastructure. This view was expressed most clearly by William Lynn III, who served as Deputy Secretary of Defense from February 2009 to October 2011 and who led efforts to create the U.S. Cyber Command and to develop the first *Department of Defense Strategy for Operating in Cyberspace*. In a 2010 article for *Foreign Affairs*, Lynn explained that

> Although the threat to intellectual property is less dramatic than the threat to critical national infrastructure, it may be the most significant cyber threat that the United States will face over the long term. Every year, an amount of intellectual property many times larger than all the intellectual property contained in the Library of Congress is stolen from networks maintained by U.S. businesses, universities, and government agencies. As military strength ultimately depends on economic vitality, sustained intellectual property losses could erode both the United States' military effectiveness and its competitiveness in the global economy.
>
> (Lynn III, 2010)

Like the CSIS report and subsequent reports and statements from the Obama administration, Lynn framed threats to critical infrastructure as still in the future and of uncertain potential impacts. Although such threats were still cause for concern because they "could have an impact analogous to physical hostilities," he noted repeatedly that "the vast majority of malicious cyber activity today does not cross this threshold" and that, therefore, current attention should focus first and foremost on informational threats (Lynn III, 2011).

Former Commander of U.S. Cyber Command, Gen. Keith Alexander, echoed this framing in explaining the *raison d'etre* and mission of his command. He explained that there is enough evidence "to be concerned about the potential effects of an actual attack" on critical infrastructure but admitted that "no one has seriously attacked these yet" (House Committee on Armed Services, 2010: 7–8). Thus, while serious threats to critical infrastructure have yet to be realized, "economic espionage for commercial and technological advantage is an everyday event" with impacts that "can take on hitherto unimaginable scale; a conqueror once had to capture a city before his army could loot it" (House Committee on Armed Services, 2010: 4).

With the dominant perceptions of the primary objects and impacts of cyber threats shifting from critical infrastructure and physical impacts to intellectual property, government data, monetary loss, and economic competitiveness, it might seem odd that the primary response measure taken by the United States thus far has been the formation of a military command. But by thoroughly linking economic competitiveness with military effectiveness and national security, Lynn and Alexander were able to turn informational objects and impacts, often in the private sector, into matters of military concern. Because private defense contractors have been among those companies that have suffered cyberattacks leading to losses of intellectual property and sensitive weapons-related information, Lynn proposed that "policymakers need to consider . . . applying the National Security Agency's defense capabilities beyond the '.gov' domain" and "look for innovative ways to use the military's cyberdefense capabilities to protect the defense industry" (Lynn III, 2010). Gen. Alexander agreed. Though he admitted that theft of intellectual property is crime, not war, and therefore "belongs more properly in law enforcement than military channels," he justified military involvement nonetheless by arguing that "when a prime target of such crime is our defense industrial base, we in the Department of Defense have a role to play in the response" (Alexander, 2011: 6).

This focus on private intellectual property, government data, and economic competitiveness was a direct driver of the United States' policy response during the Obama administration. For example, Gen. Alexander explained that it was not a major cyber incident involving critical infrastructure that led to the creation of U.S. Cyber Command, but rather, a 2008 breach of military computer networks that involved infected thumb drives (Alexander, 2010). Finally, and most importantly, Lynn's identification of intellectual property theft as the "less dramatic" but "most significant cyber threat" in his 2010 *Foreign Affairs* article was repeated almost verbatim in the July 2011 *Department of Defense Strategy for Operating in Cyberspace* in the section of the strategy that described current threats and the need for action by DOD (Department of Defense, 2011: 4).

Informational cyber threats continued to predominate even after the Director of National Intelligence (DNI) identified cyber threats in 2013 as the number one strategic threat to the United States, placing it above terrorism for the first time since September 2001. For example, in his two cyber threat assessments to Congress in 2015, the DNI identified the "costs" of the cyber threat to the United States by referencing a series of data breaches at private companies and government agencies. No critical infrastructure attacks were listed. Though the assessment pointed to reports from private sector cybersecurity researchers indicating a growing capability on the part of "unspecified Russian cyber actors" to attack critical infrastructure systems, no actual attacks were mentioned. In contrast, the DNI's assessment, in addition to hinting at the danger of cyber-enabled propaganda operations and attacks on the integrity of information, called out China specifically for ongoing "economic espionage against US companies." This framing of the problem was echoed in the updated, 2015 *DoD Cyber Strategy*. Again, while the strategy identified cyberattacks against critical infrastructure as a possibility, the document highlighted theft of "operational information and intellectual property from a range of U.S. government and commercial entities" (Department of Defense, 2015: 10), in particular at the hands of China, as a threat to "U.S. competitiveness," "strategic and technological advantage" (Department of Defense, 2015: 9–10). As we will see in later chapters, this shift towards recognition of information, to include public perceptions and opinion broadly, continued through 2016 and in the years since.

The persistent rhetoric of cyber-doom

Just as the shift in threat perceptions is not a new development, neither is the divergence between rhetoric and reality mentioned earlier. At the same time that information came to replace infrastructure as the primary object of perceived cyber threats, we can observe news media, current and former government officials, industry experts, and others deploying rhetoric of cyber-doom out of all proportion to more realistic assessments of the threat. In response, a number of observers have speculated about the persistence of such rhetoric and whether it poses a threat in and of itself.

Examples of cyber-doom rhetoric were abundant during the Obama years. In 2008, a report from the Hoover Institution warned of so-called eWMDs (Kelly and Almann, 2008). In 2009, Amit Yoran, former head of the Department of Homeland Security's National Cyber Security Division, claimed that a "cyber-9/11" had already occurred, "but it's happened slowly so we don't see it." As evidence, he pointed to the 2007 cyberattacks on Estonia, as well as other incidents in which the computer systems of government agencies or contractors had been infiltrated and sensitive information stolen (Singel, 2009). These statements echoed the 2007 comments by the speaker of the Estonian parliament, Ene Ergma, who said, "When I look at a nuclear explosion, and the explosion that happened in our country in May, I see the same thing" (Poulsen, 2007). Also in 2009, John Arquilla, one of the first to theorize cyber war in the 1990s (Arquilla and Ronfeldt, 1997), warned of "a grave and growing capacity for crippling our tech-dependent society" and has said that a "cyber 9/11" is matter of if not when (Arquilla, 2009).

The year 2010 was a particularly good year for cyber-doom rhetoric. In February, *CNN* aired *Cyber Shockwave*, a televised wargame in which former government officials battled a computer worm spreading among cell phones that eventually led to serious disruptions of critical infrastructures (Gaylord, 2010). In their book, *Cyber War: The Next Threat to National Security and What to Do about It*, Richard Clarke and Robert Knake (2010: 64–68) presented a fictional scenario in which a cyberattack destroyed all U.S. infrastructure in only fifteen minutes, killing thousands and wreaking unprecedented destruction on U.S. cities. More ominous still were the warnings from Mike McConnell, former Director of the National Security Agency, who claimed that the United States was already in an ongoing cyber war (McConnell, 2010) and even predicted that a cyberattack could surpass the impacts of 9/11 "by an order of magnitude" (The Atlantic, 2010c). Some went even further still, comparing the impacts of prospective cyberattacks to the 2004 Indian Ocean tsunami that killed roughly a quarter million people and caused widespread physical destruction in five countries (Meyer, 2010); warning that a cyberattack could have the same impact as a "well-placed bomb" (FoxNews.com, 2010a); suggesting that cyberattack could pose an "existential threat" to the United States (FoxNews.com, 2010a); and even offering the possibility that cyberattack threatens all of "global civilization" (Adhikari, 2009).

In the years since, some have used events like the financial crisis (Rothkopf, 2011b) and even the 2011 Japanese earthquake, tsunami, and Fukushima nuclear disaster, to warn of the potential impacts of cyberattacks (Rothkopf, 2011a). In 2012, while Secretary of Defense Leon Panetta was warning of a "cyber Pearl Harbor," Secretary of Homeland Security Janet Napolitano pointed to the devastation that Super Storm Sandy caused to the East Coast of the United States to warn of the potential impacts of a cyberattack (Martinez, 2012). Finally, as mentioned earlier, policy makers continue to warn of potential "cyber 9/11s" or "cyber Pearl Harbors."

Long before DNI Clapper told Congress that such cyber Armageddon scenarios were unlikely, others were pointing out that the reality of cyberattacks was far different from the rhetoric of cyber-doom. We have not seen anything close to the kinds of scenarios outlined by Yoran, Ergma, Clarke and Knake, *CNN*, and so many others. For example, as early as 2007, terrorism expert Michael Stohl noted that the cyber-doom scenarios offered for years by cybersecurity proponents had not been realized and questioned whether they were possible at all (Stohl, 2007). Such doubts seem reasonable when we recall that the real 9/11 attacks killed thousands and caused massive physical destruction, something even the most effective of cyber-kinetic attacks have yet to approach. Even with this level of destruction and disruption, the real 9/11 attacks, which had the financial heart of the nation as one of their targets, did not lead to the kind of long-term collapse of the U.S. economy often hypothesized as a possible impact of cyberattack. Instead, it was decades of bad mortgages that caused a global, financial meltdown in 2008.

Nor did the cyberattacks on Estonia approximate what happened on 9/11 as Yoran has claimed, and certainly not nuclear warfare as Ergma has claimed. In fact, a scientist at the NATO Co-operative Cyber Defence Centre of Excellence, which was established in Tallinn, Estonia, in response to the 2007 cyberattacks, wrote that the immediate impacts of those attacks were "minimal" or "nonexistent," and that "no critical services were permanently affected" (Ottis, 2010: 72). Similarly, computer security expert Bruce Schneier said of the Estonia cyberattacks, "It's kind of like an invading army coming into your country and then getting in line at the motor vehicles bureau so you can't renew your drivers license. It's not really what I think of when I think of war" (Schneier, 2010).

Not only have past incidents of cyberattack not lived up to the cyber-doom hype, there is doubt that they have the capability to do so in the future either. Though they warn of serious and ongoing threats to cybersecurity, the authors of a 2011 Organization for Economic Cooperation and Development (OECD) report assessed that cyberattacks do not have the capability to cause impacts on par with "a further failure of the global financial system, large-scale pandemics, escape of toxic substances resulting in wide-spread long-term pollution, and long-term weather or volcanic conditions inhibiting transport links across key intercontinental routes" (Sommer and Brown, 2011: 6). Similarly, though Myriam Dunn Cavelty, a cybersecurity expert at the Swiss Federal Institute of Technology, has argued that serious cyber threats do exist, nonetheless, she has said that cyber-doom scenarios are about "as likely to happen as a landing of alien spaceships" (Dunn Cavelty, 2011). Thomas Rid, professor of Strategic Studies at Johns Hopkins University's School of Advanced International Studies, declared flatly, "Cyberwar will not take place" (Rid, 2011). Finally, the largest empirical study to date of cyberattacks between rivals showed that the vast majority of attacks were of a low-level, non-destructive nature. The authors, therefore, explicitly rejected the use of worst-case scenarios to frame the policy debate on cybersecurity (Valeriano and Maness, 2015). More recent work by these and other scholars shows that most attacks continue in this vein, with Russia seen as exemplary of the growing use of cyber capabilities in support of political warfare as opposed to strikes on critical infrastructure (Valeriano et al., 2018; Kostyuk and Zhukov, 2019; Jensen et al., 2019; Valeriano and Jensen, 2019).

In response, critics have asked why there is such a divergence between cyber-doom rhetoric and the reality of actual cyber-conflict (Stohl, 2007; Weimann, 2008: 42). One possibility is that cyber threats do not exist and that cyber-doom is pure scaremongering. For example, some have noted that not only has the official narrative about who threatens what, how, and with what potential impact shifted over time, but it has done so with very little evidence provided to support the claims being made, thus raising the possibility that cyber threats are a mere fiction (Bendrath, 2001, 2003; Walt, 2010). Some of these critics portray the various think tanks, security firms, defense contractors, and government leaders who trumpet the problem of cyberattacks as self-interested ideologues who promote unrealistic portrayals of cyber threats for the financial benefit of an emerging "cyber-industrial complex" (Greenwald, 2010; Brito and Watkins, 2011; Blunden and Cheung, 2014).

Others take cyber threats more seriously, but still wonder about the persistence of the threat inflation found in cyber-doom rhetoric. They link cyber-doom rhetoric to larger social, cultural, and political currents. For example, some point to the fact that fears of cyberterrorism and cyber war combine a number of long-standing human fears, including fear of terrorism (especially since 9/11), fear of the unknown, and fear of new technologies (Stohl, 2007; Weimann, 2008: 42; Embar-Seddon, 2002: 1034), thus making hypothetical cyber-doom a particularly salient fear. Similarly, Dunn Cavelty has explained the persistence of cyber-doom rhetoric by linking cybersecurity threats to the emergence of a whole host of new, more uncertain, and ambiguous threats in the post-Cold War period. "[T]hat they are hyped," she has written, "is in fact just a side effect of their nature" (Dunn Cavelty, 2008: 6).

Nonetheless, as we saw earlier, Dunn Cavelty warns against placing too much stock in "hysterical doomsday scenarios" when formulating cybersecurity policy. This is because there are several potential dangers of relying too heavily on such worst-case thinking. Framing cyber threats in such terms invites a militarized response that may be ineffective and even counterproductive for dealing with the cyber threats we do face (Lewis, 2010). In the first case, it is not at all clear that the military is the appropriate institution for dealing effectively with the kind of broad cyber threat to private intellectual property and personal information identified by DNI Clapper and his predecessors. More concerning, however, is the possibility that the types of policies and responses that worst-case thinking promotes are actually counterproductive. Militarized cybersecurity policies by the United States could undermine its own policy of promoting Internet freedom around the world. In an interconnected environment such as cyberspace, the kinds of offensive actions often contemplated or, in some cases already undertaken, can "blow back" onto the party who initiated those actions, leading to unintended, negative consequences (Dunn Cavelty, 2008: 143; Lawson, 2015; Dunn Cavelty and Van Der Vlugt, 2015). In other cases, such framing might encourage defensive actions that are ineffective of even counterproductive (Ball et al., 2013; Gallagher and Greenwald, 2014; Schneier, 2014). There also exists the possibility that worst-case, cyber-doom thinking could distract from and lead to a sense of complacency about the more mundane but realistic cyber threats we do face (Debrix, 2001: 156; Lewis, 2010: 4; Lawson, 2012). Finally, some worry that worst-case, cyber-doom thinking and the militarized responses it promotes could end up as a self-fulfilling prophecy, lead to conflict escalation where non-physical cyberattacks escalate to physical warfare, or lead to the kinds of preventive war scenarios witnessed in the 2003 U.S. invasion of Iraq (Furedi, 2009; Thierer, 2013; Blunden and Cheung, 2014; Valeriano and Maness, 2015).

A critical appraisal of cyber-doom

Given the persistence of cyber-doom rhetoric, including speculation about the reasons for its persistence and its possible effects on our ability to respond to legitimate cybersecurity problems, this book investigates cyber-doom rhetoric specifically and in detail. It traces the emergence, persistence, and influence of cyber-doom rhetoric in U.S. public discourse about threats to and through cyberspace. It provides a critical appraisal that describes not only the ways that various actors have deployed cyber-doom rhetoric, but also various conditions that have allowed for the emergence and persistence of cyber-doom rhetoric and its influence in shaping the U.S. Information-Age security imaginary. To accomplish this task, this book identifies a number of variations of cyber-doom rhetoric and their deployment in various contexts and by various actors in U.S. public discourse about cybersecurity. It has sought to understand cyber-doom rhetoric in relation to longer-term, historical concerns in the West about Information-Age security in particular and the relationships among technology and society more generally. Finally it has sought to assess the implications of cyber-doom rhetoric by tracing the vision of society, technology, and security that it articulates. In short, this book investigates how, why, and with what potential impacts cyber-doom rhetoric is articulated as an element within, but also helps to articulate, the U.S. Information-Age security imaginary.

The Information-Age security imaginary is that portion of the social imaginary focused on the supposedly unique security challenges of the Information Age. The concept of the "social imaginary" refers to "the ways people imagine their social existence, how they fit together with others, how things go on between them and their fellows, the expectations that are normally met, and the deeper normative notions and images that underlie these expectations," which together form the foundations for "common understanding that makes possible common practices and a widely shared legitimacy." Imaginaries are "both factual and normative," entailing not only "a sense of how things usually go," but also "an idea of how they ought to go" (Taylor, 2004: 23–4). The "security imaginary" is "that part of the social imaginary that deals with the understanding of the security world and in turn makes security practices possible" (Pretorius, 2008: 117). It is about how "security and insecurity (or threat) . . . are constructed through the fixing of meanings to things, an identity to 'the self' and others, and the relationships that are thus instituted" (Pretorius, 2008: 100).

Social and security imaginaries are articulations of heterogeneous elements that serve to describe and enact a vision of the world and their creators' places within it. "Articulation" here refers to a phenomenon, a theory, and a method. It is a method of analyzing the way that "discursive structures constitute and organize social relations" that begins with the observation that the formation of those discursive structures is the "result of an articulatory practice" (DeLuca, 1999: 335). Articulation as practice involves enunciating elements (e.g. beliefs, values, individuals, organizations, technologies, practices, discourses, and more) and then linking those elements into a "unity," which often has the effect of empowering certain ways of seeing, acting, and being while disempowering or constraining others. Security imaginaries as articulations are contingent and occur within and help to constitute particular historical conjunctures (Laclau and Mouffe, 1985: 105, 113; Grossberg, 1986: 53; DeLuca, 1999: 335; Slack, 2006: 225). Scholars who have studied security imaginaries have deployed articulation to trace their emergence. This has included studies investigating the Cuban missile crisis (Weldes, 1996, 1999), the phenomenon of military isomorphism (Pretorius, 2008), U.S. military theories of "perception management" and "information warfare" (Brunner and Dunn Cavelty, 2009), the Bush Administration's case for ballistic missile defense (Sikka, 2008), controversy in the U.S. military over the use of social media (Lawson, 2013b), and the development of industrial- and Information-Age theories of warfare (Lawson, 2011, 2014).

Drawing heavily from the linguistics of Ferdinand de Saussure and the post-structuralist philosophy of Jacques Derrida, Ernesto Laclau and Chantal Mouffe introduced the concept of "antagonism" as a way of accounting for change in articulations over time. Starting from the premise that articulations of any kind, including the social imaginary, are systems of differences like language, Laclau and Mouffe argue that the antagonism represents the fact that such systems are never entirely stable. Thus, the "antagonism as the negation of a given order is, quite simply, the limit of that order," which in turn arises from within the system itself "as something subverting it, destroying its ambition to constitute a full presence" (Laclau and Mouffe, 1985: 126–7). While there are no essential, *a priori* antagonisms (DeLuca, 1999: 336; Laclau and Mouffe, 1985: 127) – that is, no particular forms of antagonism are inevitable – nonetheless, the emergence of antagonisms is a "natural" result of the fact that no articulation can ever be total or complete.

In recent decades, social imaginaries in general, and security imaginaries in particular, have increasingly been articulated in response to rapid changes in science and technology (Marcus, 1995: 3–4) and the perceived challenges of navigating "complex technical, social, and political-economic systems" (Fortun and Fortun, 2005: 44). Not only can the introduction of disruptive new technologies serve as a form of antagonism (Lawson, 2013b), but even the general perception of technoscientific change that seems to "run ahead" of society, economy, law, or cultural norms and values, can serve as an antagonism leading to shifts in social and security imaginaries. Indeed, perceptions of intercalated technoscientific change as antagonism are a defining characteristic of the U.S. Information-Age security imaginary (Lawson, 2011). For professionals of security in the United States, the Information Age, with its attendant changes in technology and society, is imagined to have ushered in a newly volatile, uncertain, complex, ambiguous, and dangerous world that challenges traditional practices of knowing, acting, and being (Stiehm, 2002; Lawson, 2014). Thus, concern with threats to and through the global network of interconnected telecommunication infrastructures upon which so much of information society and economy depend (i.e. the cyberspace "domain") has emerged as a prominent feature of this Information-Age security imaginary.

An important caveat is warranted at this point. The use of the word "imaginary" does not mean "unreal." Imaginaries are made, but not made up (Bonditti et al., 2015: 170). Nonetheless, imagination can and does play an important role in the articulation of imaginaries. Theorists of articulation have noted that language and rhetoric, like the use of particular metaphors, can serve as important elements in articulations and are therefore valuable indicators of dominant structures of meaning (Makus, 1990: 504). Beyond the use of metaphor, however, fiction, fantasy, and hypothetical scenarios have a long history in the formation of security imaginaries (Clarke, 1966, 1979, 1995, 1999; Ghamari-Tabrizi, 2000; Franklin, 2001; Ghamari-Tabrizi, 2004; Muller, 2008; Lawson, 2011, 2014). This book uses articulation to examine cyber-doom rhetoric, including its dominant metaphors of war and disaster and its use of hypothetical scenarios, as constitutive elements of the U.S. Information-Age security imaginary.

To trace the role of cyber-doom rhetoric in the articulation of the U.S. Information-Age security imaginary, this book relies on a wide variety of both elite and non-elite sources, from official policy documents and statements, to journalistic accounts, to fiction, and more. These materials were collected over a period of a decade of closely following the U.S. cybersecurity debate following Altheide's "discourse tracking" methodology, an ethnographic approach to discourse analysis that takes advantage of the ability that the Internet and electronic databases allow for searching, collecting, and analyzing fragments of public discourse over long periods of time (Altheide and Schneider, 2013: 8; See also Altheide, 2000). Examining such a wide variety of sources is necessary to understand the complex, intertextual, and iterative construction of security imaginaries (Hansen, 2006: 55; Pretorius, 2008: 103, 105). As Salter notes, "There are multiple discursive and technocratic fronts on which the competition for attention and resources is played out" (Salter, 2011: 117). There is not one speaker and one audience involved in the production of security imaginaries. Rather, the security imaginary is a product of "many elite and public authors" who marshal "technical, political, and cultural arguments" (Hasian et al., 2015: 2). Of course, we should not ignore official statements and policy texts because, as Taylor notes, "it often happens that what start off as theories held by a few people come to infiltrate the social imaginary, first of elites, perhaps, and then of the whole society" (Taylor, 2004: 24). Nonetheless, Hansen cautions that we should not treat such texts "as entities standing separately from wider societal discourses but as entities located within a larger textual web; a web that both includes and goes beyond other policy texts, into journalism, academic writing, popular non-fiction, and, potentially, even fiction" (Hansen, 2006: 55). Ultimately, an examination of a wide variety of texts is necessary because cyber-doom rhetoric, like rhetoric more generally, "is not some static entity but a mobile and protean force, made of fragmentary flows and cultural signifiers that circulate both within, and across, porous spheres and boundaries" (Hasian et al., 2015: 7).

As articulations, security imaginaries are not entirely unique. They are conditioned and shaped by a number of factors, including the wider social imaginary of their time, prior articulations of the security imaginary, and the opportunities and constraints offered by the process of articulation itself (Lawson, 2014: 11–12). As such, this book begins by locating the emergence of cyber-doom rhetoric as an element within a broader Information-Age security imaginary that emerged at the end of the Cold War. Rapid developments in information technology, alongside massive social, political, and economic changes on a global scale, resulted in a great deal of anxiety about the implications of these changes for the security of societies in the Information Age. As noted earlier, even as some have criticized cyber-doom rhetoric and official assessments of cyber threats have shifted, cyber-doom rhetoric has persisted as a vestige of our early Information-Age security imaginary.

This book will also argue, however, that cyber-doom rhetoric persists because it is conditioned by and reflective of several broader, historical traditions in Western societies. These include a fear of technology-out-of-control and, in particular, concern about the supposed fragility of modern societies marked by complex interdependencies and reliance on large technological systems for their functioning. Such concerns have been accompanied by a tradition of using fiction and hypothetical, worst-case thinking to imagine and plan for the future. Cyber-doom rhetoric is an incarnation of this tradition. Thus, while cyber-doom scenarios have not yet been realized, the repetition of these stories about a frightening potential future is itself a real social phenomenon now, in the present, potentially producing real effects in the present and the future.

That is why it is important to understand cyber-doom rhetoric, including its variations, uses, and the vision it articulates. In later chapters, I will argue that cyber-doom rhetoric is not primarily deployed to diagnose or describe real cyber threats. Rather, it is a fear appeal that relies on the appropriation of other non-cyber disasters, the exaggeration of more mundane cyber incidents, or the use of pure fiction, as a motivational tool or call to action. But in attempting to raise awareness of and motivate a response to cybersecurity threats, cyber-doom rhetoric also specifies, defines, and links conceptions of society, economy, technology, governance, the state, the individual, and more, to express a vision of the current state of information society, to imagine a frightening potential future, and to suggest possible responses here and now to reform society and avoid the worst.

It is this vision and the responses that it implies that raise concern about the potentially negative impacts of cyber-doom rhetoric. Cyber-doom rhetoric articulates a vision of a fragile society and economy dependent on technology increasingly believed to be out of our control, a society and economy always already on the verge of collapse. As mentioned earlier, pure imagination, fictions, hypothetical scenarios, and metaphors or analogies to the most frightening past experiences are prominent techniques used to produce "knowledge" of cyber-doom. In extreme cases, cyber-doom rhetoric can even be seen as against empirical knowledge, relying instead on fear and "you never knowism" to promote preventive action to avoid the worst (Furedi, 2009: 214–15). In such a vision, there is no time for facts or knowledge; we must act now (Elmer and Opel, 2008; Furedi, 2009: 208).

The relationships among the state and individuals in society are also implicated in the articulation of the cyber-doom vision. I will argue later in the book that cyber-doom rhetoric can encourage authoritarian and/or anti-social relations of power and ethics. In some variants, the individual is helpless and in need of a reassertion of centralized and militarized state power that governs through anticipating and preempting the worst possible futures. This marks a move towards authoritarianism and concomitant threats to individual rights and liberties. In other variants of cyber-doom rhetoric, the state is rendered irrelevant and only a super-empowered, responsibilized individual, the neoliberal resurrection of the rugged individual of the American frontier, can weather the storm of cyber-doom and emerge to rebuild society. Still other rehearsals of cyber-doom offer an uneasy combination of the two, a sort of "convergence security" that can take the form of vigilantism in support of the national security state (Lawson and Gehl, 2011).

The methodological approach described earlier is broadly genealogical and deconstructive, meaning that the book begins by engaging in a "second-order observation" of the "problematization" of cyberspace (Luhmann, 1993, 2000; Collier et al., 2004; Bonditti et al., 2015). This means that our investigation focuses first and foremost on the conditions of possibility that have allowed cyber-doom rhetoric to become a taken-for-granted feature of a U.S. discourse that has constructed cyberspace as a security problem. We begin not by asking "first-order" questions about whether cyber-doom rhetoric is an accurate reflection of real cyber threats, but rather, asking "second-order" questions about how and why cyber-doom rhetoric has emerged in the first place, the role it has played, and why it has persisted even in the face of assessments and events that would seem to undermine its credibility.

But this book also takes seriously the idea that such approaches need not, and should not, only mean standing on the outside, describing and criticizing, while ignoring engagement with "first-order" or normative questions (Caputo, 1997: 5–6, 9; Floyd, 2011). Methodological approaches that engage in second-order observation, that "undermine naturalized assumptions," and aim to "reveal the contingent power relations behind them," are a powerful first step towards "mak[ing] new forms of freedom, change, and creativity possible" and "open[ing] up the range of possibilities for thinking and acting" (Bonditti et al., 2015: 163, 171). Therefore, in addition to examining the emergence and persistence of cyber-doom rhetoric, this book also assesses how realistic the fears expressed in cyber-doom rhetoric are and offers guidance for how we might development alternatives for thinking more productively about the cybersecurity challenges that we face.

Ultimately, this book argues that fears of social, economic, or even civilizational chaos and collapse, which are prominent fears expressed in cyber-doom rhetoric, are unwarranted. It agrees with a diverse collection of scholars and practitioners of security who recognize the importance of language, including metaphor, analogy, and narrative, for appropriately framing and then responding to problems. Thus, it will make the case for thinking about cybersecurity threats using other metaphors and analogies, both military and non-military, such as biological warfare, counterinsurgency, the human immune system, and complex adaptive systems. The implications of this analysis are that our understandings of the cybersecurity challenges we face must be based more on empirical evidence and less on the fears and fictions of cyber-doom. In the end, making this shift means that our responses to cybersecurity challenges should focus less on those that seek preemption, centralization, and militarization and more on those that are guided by principles of resilience, decentralization, and self-organization.

Summary of chapters

The chapters of this book are split evenly between examining the second- and first-order questions related to cyber-doom rhetoric. The first three chapters focus on second-order questions related to the emergence and persistence of cyber-doom rhetoric. The remaining chapters address first-order questions related to the real-ism, effects, and alternatives to cyber-doom rhetoric for framing cybersecurity challenges. As noted earlier, cyber-doom rhetoric often proceeds through the use of hypotheticals and, in the most extreme cases, can promote a form of preventive action based in "you never knowism," that is, acting out of fear to prevent the worst before all the facts are in. However, there is a great deal of existing research from a number of fields that can help us make sense of the potential impacts of cybersecurity threats. We do not need to rely on fear and fiction to guide our way. As a result, the chapters that follow look for insights in scholarly work from a number of disciplines. The remainder of this chapter provides an overview of these chapters and their respective sources and arguments.

This introductory chapter has begun the second-order observation of cyber-doom rhetoric by defining the phenomenon and placing it within the immediate context of a shift in cyber threat perceptions during the presidency of Barack Obama. Chapter 2 continues this work by looking in more detail at the early history and emergence of cyber-doom at the end of the Cold War, as well as the variations of its use by different speakers in different contexts, including industry experts, journalists, policymakers, and entertainment media. It argues that while cyber-doom may have initially been a reflection of real concerns about cyber threats, it now functions primarily as a form of argument known as a fear appeal, which is used to raise awareness of, and motivate a response to, cybersecurity challenges that are much more mundane than those envisioned in cyber-doom rhetoric.

Chapter 3 explores a number of broader historical and social conditions that have allowed for the emergence and persistence of cyber-doom rhetoric. This chapter looks to work from the history of technology, military history, sociology, psychology, and communication to argue that cyber-doom rhetoric is the latest manifestation of fears of technology-out-of-control in Western societies. From the work of historians, we know that this has included fears about electrification and early communications technologies like radio and telegraph during the industrial age. Western militaries, including the United States military, have not been immune to these fears and have, as a result, often treated new technologies, including information and communication technologies, with ambivalence and even fear. Along with the technological advancements of the last century have come repeated bouts of what psychologists call technophobia and sociologists call technopanic. Fear of cyber-doom is but the latest incarnation of these fears.

In Chapter 4, the focus of our investigation shifts to first-order questions. This chapter assesses how realistic the fears expressed in cyber-doom rhetoric are. This chapter draws from work in history, sociology, and recent experiences with cyberattack to argue that cyberattacks against critical infrastructure like the ones contemplated in cyber-doom scenarios, though not out of the question, are nonetheless unlikely to cause the social and economic panic, chaos, and collapse often envisioned as the ultimate effects of such scenarios. This assessment is based on an examination of the history of strategic bombing and failures of critical infrastructures like mass blackouts; disasters like the attacks of September 11, 2001, and Hurricane Katrina; the findings of research in disaster sociology; and the actual effects of the most prominent, large-scale cyberattacks like those against Georgia and Estonia, the Stuxnet attack against Iran, and cyber operations during the Ukraine conflict.

Chapter 5 explores a number of potentially negative implications of cyber-doom rhetoric for framing and responding to cybersecurity challenges. It draws from work in psychology, sociology, and communication to argue that language, including metaphor and analogy, is important for appropriately framing and responding to problems. It argues that cyber-doom fear appeals are potentially dangerous because they can promote a form of "you never knowism" along with a number of counterproductive responses to cyber threats. The effects of cyber-doom fear appeals can involve defining cyber threats too narrowly, focusing on the worst but unlikely threats while ignoring the less dramatic but real threats that we do face, taking actions in response to hypothetical threats that are inappropriate or counterproductive in response to actual threats, or causing a sense of fatalism and paralysis among citizens and policymakers in the face of frightening scenarios to which they see no efficacious response.

Chapter 6 provides a critical appraisal of the post-2016 debate in the United States over how to understand and talk about the Russian operations and what they portend for the future of international cyber conflict. It describes the terminological and conceptual confusion since 2016, with some claiming that Russian operations were a case of cyber war, while others turn instead to terms and concepts like "information warfare," "hybrid warfare," and "political warfare." This chapter argues that the debate about terms and concepts is itself an indicator of a serious challenge to the dominance of cyber-doom rhetoric. It evaluates the historical meaning of each term and its fit with what we know of Russia's 2016 cyber operations, arguing that political warfare provides the most apt starting point for thinking about the future of cyber operations in international conflict.

Finally, Chapter 7 concludes the book by reviewing its main arguments in an effort to offer some lessons learned for improving the way we think and speak about cybersecurity challenges. It makes the case for exploring a range of alternative analogies and metaphors beyond the usual suspects and beyond those currently ascendant in the U.S. cybersecurity debate. These could include biological warfare, counterinsurgency, the human immune system, and complex adaptive systems. From this perspective, it argues that we must define cybersecurity more broadly, draw from a wider range of knowledges and expertise in thinking about and formulating responses to cybersecurity challenges, and in all cases prioritize empirical evidence over hypothetical scenarios. Finally, it argues that our responses to cybersecurity challenges should focus less on those that seek preemption, centralization, and militarization and more on those that are guided by principles of resilience, decentralization, and self-organization. In these ways, we can begin to move beyond the inaccurate and potentially counterproductive rhetoric of cyber-doom.

References

Adhikari, R. (2009) "Civilization's High Stakes Cyber-struggle: Q&A with Gen. Wesley Clark (ret.)," *TechNewsWorld*, 2 December 2009. Online. Available: <www.technewsworld. com/story/Civilizations-High-Stakes-Cyber-Struggle-QA-With-Gen-Wesley-Clark-ret-68787.html?wlc=1259861126&wlc=1259938168&wlc=1290975140> (accessed 2 December 2009).

Alexander, G.K. (2010) *U.S. Cybersecurity Policy and the Role of U.S. CYBERCOM*, Presented at Center for Strategic and International Studies, 3 June 2010.

———. (2011) "Building a New Command in Cyberspace," *Strategic Studies Quarterly*, 5, 2: 3–12.

Altheide, D.L. (2000) "Tracking Discourse and Qualitative Document Analysis," *Poetics*, 27, 4: 287–99. doi:10.1016/S0304-422X(00)00005-X.

Altheide, D.L. and Schneider, C.J. (2013) *Qualitative Media Analysis*, Los Angeles: Sage Publications. doi:10.4135/9781452270043.

Arquilla, J. (2009) "Click, Click . . . Counting Down to Cyber 9/11," *San Francisco Chronicle*, 26 July 2009, E2.

Arquilla, J. and Ronfeldt, D. (1997) "Cyberwar Is Coming!," in Arquilla, J. and Ronfeldt, D. (eds) *In Athena's Camp: Preparing for Conflict in the Information Age*, Santa Monica, CA: RAND, pp. 24–60.

The Atlantic. (2010c) "Fmr. Intelligence Director: New Cyberattack May Be Worse Than 9/11," *The Atlantic*, 30 September 2010c. Online. Available: <www.theatlantic.com/politics/archive/2010/09/fmr-intelligence-director-new-cyberattack-may-be-worse-than-9-11/63849/> (accessed 30 September 2010c).

Ball, J., Borger, J. and Greenwald, G. (2013) "Revealed: How U.S. and U.K. Spy Agencies Defeat Internet Privacy and Security," *The Guardian*, 6 September 2013. Online. Available: <www.theguardian.com/world/2013/sep/05/nsa-gchq-encryption-codes-security> (accessed 6 September 2013).

Bendrath, R. (2001) "The Cyberwar Debate: Perception and Politics in Us Critical Infrastructure Protection," *Information & Security: An International Journal*, 7: 80–103. doi:10.11610/isij.0705.

———. (2003) "The American Cyber-Angst and the Real World—Any Link," in Latham, R. (ed) *Bombs and Bandwidth: The Emerging Relationship Between Information Technology and Security*, New York: The Free Press, pp. 49–73.

Blank, S. (2008) "Web War I: Is Europe's First Information War a New Kind of War?," *Comparative Strategy*, 27, 3: 227–47. doi:10.1080/01495930802185312.

Blunden, W. and Cheung, V. (2014) *Behold a Pale Farce: Cyberwar, Threat Inflation, & The Malware-Industrial Complex*, Waterville, OR: Trine Day.

Bonditti, P., et al. (2015) "Genealogy," in Aradau, C. et al. (eds) *Critical Security Methods: New Frameworks for Analysis*, London: Routledge, pp. 159–88. doi:10.4324/9781315881549.

Brito, J. and Watkins, T. (2011) *Loving the Cyber Bomb? The Dangers of Threat Inflation in Cybersecurity Policy*, Mercatus Center Working Paper No. 11–24, April.

Brunner, E. and Dunn Cavelty, M. (2009) "The Formation of in-Formation By the Us Military: Articulation and Enactment of Informatic Threat Imaginaries on the Immaterial Battlefield of Perception," *Cambridge Review of International Affairs*, 22, 4: 629–46. doi:10.1080/09557570903325454.

Bumgarner, J. and Borg, S. (2009) *Overview By the US-CCU of the Cyber Campaign Against Georgia in August of 2008*, US-CCU Special Report, August.

Caputo, J.D. (1997) *Deconstruction in a Nutshell: A Conversation With Jacques Derrida (Perspectives in Continental Philosophy)*, New York: Fordham University Press.

Carman, A. (2015) "OPM Breaches More Serious to National Security Than 9/11, Congresswoman Argues During Hearing," *SC Magazine*, 16 June 2015. Online. Available: <www.scmagazine.com/house-committee-on-oversight-and-government-reform-hosts-hearing-on-data-breaches/article/421052/> (accessed 16 June 2015).

Carr, H. (2017) "Carr: 'Russian' to Conclusions, Dems' Focus Is on Gossip," *Boston Herald*, 14 June 2017. Online. Available: <www.bostonherald.com/news/columnists/howie_carr/2017/06/carr_russian_to_conclusions_dems_focus_is_on_gossip> (accessed 14 June 2017).

Chang, L. and Osborne, M. (2017) "Meddling," *RawData*, 18 May 2017. Online. Available: <https://worldview.stanford.edu/raw-data/meddling> (accessed 18 May 2017).

Clarke, I.F. (1966) *Voices Prophesying War, 1763–1984*, London: Oxford University Press.

———. (1979) *The Pattern of Expectation, 1644–2001*, New York: Basic Books.

———. (1995) *The Tale of the Next Great War, 1871–1914: Fictions of Future Warfare and of Battles Still-to-come*, Syracuse, NY: Syracuse University Press.

Clarke, L. (1999) *Mission Improbable: Using Fantasy Documents to Tame Disasters*, Chicago: University of Chicago Press.

Clarke, R.A. and Knake, R. (2010) *Cyber War: The Next Threat to National Security and What to Do About It*, New York: HarperCollins.

Clinton, H.R. (2010) *Remarks on Internet Freedom*, Presented at The Newseum, Washington, DC, 21 January 2010.

Collins, S.J. (2015) "Senator Collins Continues to Sound the Alarm on the Urgent Need to Bolster Cybersecurity," *Press Release, Office of Senator Susan Collins*, 6 August 2015. Online. Available: <www.collins.senate.gov/public/index.cfm/2015/8/senator-collins-continues-to-sound-the-alarm-on-the-urgent-need-to-bolster-cybersecurity> (accessed 6 August 2015).

Collier, S.J., Lakoff, A. and Rabinow, P. (2004) "Biosecurity: Towards an Anthropology of the Contemporary," *Anthropology Today*, 20, 5: 3–7. doi:10.1111/j.0268-540x.2004.00292.x.

Committee on Homeland Security and Governmental Affairs. (2011a) *Cyber Security and American Cyber Competitiveness Act of 2011 (S.21)*, 112th Congress, 1st Session, United States Senate, 25 January 2011a.

———. (2011b) *Cyber Security Public Awareness Act of 2011 (S.813)*, 112th Congress, 1st Session, United States Senate, 13 April 2011b.

Conway, M. (2008) "Media, Fear and the Hyperreal: The Construction of Cyberterrorism as the Ultimate Threat to Critical Infrastructures," in Dunn Cavelty, M. and Kristensen, K.S. (eds) *Securing the 'Homeland': Critical Infrastructure, Risk and (in)security*, London: Routledge, pp. 109–29.

Curran, K. (2012) "Viewpoint: James Bond Fails the Tech Test in Skyfall," *BBC News*, 2 December 2012. Online. Available: <www.bbc.com/news/technology-20555621> (accessed 2 December 2012).

Davis, J.H. (2015) "Hacking of Government Computers Exposed 21.5 Million People," *New York Times*, 9 July 2015. Online. Available: <www.nytimes.com/2015/07/10/us/office-of-personnel-management-hackers-got-data-of-millions.html> (accessed 9 July 2015).

Debrix, F. (2001) "Cyberterror and Media-Induced Fears: The Production of Emergency Culture," *Strategies*, 14, 1: 149–68. doi:10.1080/10402130120042415.

DeLuca, K. (1999) "Articulation Theory: A Discursive Grounding for Rhetorical Practice," *Philosophy & Rhetoric*, 32, 4: 334–48.

Department of Defense. (2011) *Department of Defense Strategy for Operating in Cyber-space*, 2011. Online. Available: <www.defense.gov/news/d20110714cyber.pdf> (accessed 2011).

———. (2015) *DOD Cyber Strategy*, Washington, DC: Department of Defense.

Dickinson, B. (2012) "This Time, James Bond Fights Hackers," *Information Week*, 12 November 2012. Online. Available: <www.informationweek.com/desktop/this-time-james-bond-fights-hackers/d/d-id/1107342?> (accessed 12 November 2012).

Dunn Cavelty, M. (2008) *Cyber-Security and Threat Politics: U.S. Efforts to Secure the Information Age*, New York: Routledge. doi:10.4324/9780203937419.

———. (2011) "As Likely as a Visit From E.T.," *The European*, 7 January 2011. Online. Available: <www.theeuropean-magazine.com/133-cavelty/134-cyberwar-and-cyberfear> (accessed 7 January 2011).

Dunn Cavelty, M. and Van Der Vlugt, R.A. (2015) "A Tale of Two Cities: Or How Wrong Metaphors Lead to Less Security," *Georgetown Journal of International Affairs*, Fall: 21–9.

Elmer, G. and Opel, A. (2008) *Preempting Dissent: The Politics of an Inevitable Future*, Winnipeg: Arbeiter Ring Pub.

Embar-Seddon, A. (2002) "Cyberterrorism: Are We Under Siege?," *American Behavioral Scientist*, 45, 6: 1033–43. doi:10.1177/0002764202045006007.

Epstein, Z. (2012) "How a Real-life Computer Virus Inspired the Latest James Bond Film 'Skyfall'," *BGR.com*, 8 November 2012. Online. Available: <https://bgr.com/2012/11/08/james-bond-skyfall-plot-cyberterrorism/> (accessed 8 November 2012).

Evron, G. (2008) "Battling Botnets and Online Mobs: Estonia's Defense Efforts During the Internet War," *Georgetown Journal of International Affairs*, 9, Winter/Spring: 121–6.

Floyd, R. (2011) "Can Securitization Theory Be Used in Normative Analysis? Towards a Just Securitization Theory," *Security Dialogue*, 42, 4–5: 427–39. doi:10.1177/0967010611418712.

Fortun, M. and Fortun, K. (2005) "Scientific Imaginaries and Ethical Plateaus in Contemporary U.S. Toxicology," *American Anthropologist*, 107, 1: 43–54. doi:10.1525/aa.2005.107.1.043.

FoxNews.com. (2010a) "FBI Warns Brewing Cyberwar May Have Same Impact as 'Well-placed Bomb'," *FOXNews.com*, 8 March 2010a. Online. Available: <www.foxnews.com/tech/2010/03/08/cyberwar-brewing-china-hunts-wests-intel-secrets/> (accessed 8 March 2010a).

Franklin, B. (2001) *War Stars: The Superweapon and the American Imagination*, London: Oxford University Press.

Furedi, F. (2009) "Precautionary Culture and the Rise of Possibilistic Risk Assessment," *Erasmus Law Review*, 2, 2: 197–220.

Gallagher, R. and Greenwald, G. (2014) "How the NSA Plans to Infect 'Millions' of Computers With Malware," *The Intercept*, 12 March 2014. Online. Available: <https://firstlook.org/theintercept/2014/03/12/nsa-plans-infect-millions-computers-malware/> (accessed 12 March 2014).

Gaylord, C. (2010) "Cyber Shockwave Cripples Computers Nationwide (Sorta)," *Christian Science Monitor*, 16 February 2010. Online. Available: <www.csmonitor.com/Innovation/Horizons/2010/0216/Cyber-ShockWave-cripples-computers-nationwide-sorta> (accessed 16 February 2010).

Geers, K. (ed) (2015) *Cyber War in Perspective: Russian Aggression Against Ukraine*, Tallinn, Estonia: NATO CCDCOE Publications.

Ghamari-Tabrizi, S. (2000) "Simulating the Unthinkable: Gaming Future War in the 1950s and 1960s," *Social Studies of Science*, 30, 2: 163–223. doi:10.1177/030631200030002001.

———. (2004) "The Convergence of the Pentagon and Hollywood: The Next Generation of Military Training Simulations," in Rabinowitz, L. and Geil, A. (eds) *Memory Bytes: History, Technology, and Digital Culture*, Durhan, NC: Duke University Press, pp. 150–73. doi:10.1215/9780822385691.

Graham, T. (2017) "Maddow Indulges 'The Great' Dan Rather: Russians Pulled Off a 'Cyber Pearl Harbor' In 2016," *NewsBusters*, 10 June 2017. Online. Available: <www.newsbusters.org/blogs/nb/tim-graham/2017/06/10/maddow-indulges-great-dan-rather-russians-pulled-cyber-pearl-harbor> (accessed 10 June 2017).

Greenberg, A. (2012) *This Machine Kills Secrets: How Wikileakers, Cypherpunks and Hacktivists Aim to Free the World's Information*, New York: Dutton.

Greenwald, G. (2010) "Mike McConnell, the Washpost & The Dangers of Sleazy Corporatism," *Salon.com*, 29 March 2010. Online. Available: <www.salon.com/news/opinion/glenn_greenwald/2010/03/29/mcconnell> (accessed 29 March 2010).

———. (2014) *No Place to Hide: Edward Snowden, the NSA, and the U.S. Surveillance State*, New York: Metropolitan Books.

Grossberg, L. (1986) "On Postmodernism and Articulation: An Interview With Stuart Hall," *Journal of Communication Inquiry*, 10, 2: 45–60. doi:10.1177/019685998601000204.

Hansen, L. (2006) *Security as Practice: Discourse Analysis and the Bosnian War*, New York: Routledge.

Hasian, M., Lawson, S.T. and McFarlane, M. (2015) *The Rhetorical Invention of America's National Security State*, Lanham, MA: Lexington Books.

Hertling, M. and McKew, M.K. (2018) "Putin's Attack on the U.S. Is Our Pearl Harbor," *Politico*, 16 July 2018. Online. Available: <www.politico.com/magazine/story/2018/07/16/putin-russia-trump-2016-pearl-harbor-219015> (accessed 16 July 2018).

House Committee on Armed Services. (2010) *Statement of General Keith B. Alexander Commander United States Cyber Command*, United States House of Representatives, 23 September 2010.

House Permanent Select Committee on Intelligence. (2015) *Statement for the Record: Worldwide Cyber Threats*, United States House of Representatives, 10 September 2015b.

Hubert, K. (2012) "Don't Skyfall for It," *BusinessTechnology*, 20 November 2012. Online. Available: <https://business-technology.co.uk/2012/11/dont-skyfall-for-it/> (accessed 20 November 2012).

Jensen, B., Valeriano, B. and Maness, R. (2019) "Fancy Bears and Digital Trolls: Cyber Strategy With a Russian Twist," *Journal of Strategic Studies*, 42, 2: 1–23. doi:10.1080/01402390.2018.1559152.

Kaplan, F. (2016) *Dark Territory: The Secret History of Cyber War*, New York: Simon & Schuster.

Kelly, J.J. and Almann, L. (2008) "eWMDs: The Botnet Peril," *Policy Review*, 152. Online. Available: <www.hoover.org/publications/policy-review/article/5662>.

Korns, S.W. and Kastenberg, J.E. (2008) "Georgia's Cyber Left Hook," *Parameters*, Winter: 60–76.

Kostyuk, N. and Zhukov, Y.M. (2019) "Invisible Digital Front: Can Cyber Attacks Shape Battlefield Events," *Journal of Conflict Resolution*, 63, 2: 317–47. doi:10.1177/0022002717737138.

Laclau, E. and Mouffe, C. (1985) *Hegemony and Socialist Strategy: Towards a Radical Democratic Politics*, London: Verso.

Langevin, R.J.R., et al. (2008) *Securing Cyberspace for the 44th Presidency*, Washington, DC: Center for Strategic and International Studies.

Lawson, S. (2011) "Articulation, Antagonism, and Intercalation in Western Military Imaginaries," *Security Dialogue*, 42, 1: 39–56. doi:10.1177/0967010610393775.

———. (2012) "Of Cyber Doom, Dots, and Distractions," *Forbes.com*, 16 October 2012. Online. Available: <www.forbes.com/sites/seanlawson/2012/10/16/of-cyber-doom-dots-and-distractions/> (accessed 16 October 2012).

———. (2013a) "Motivating Cybersecurity: Assessing the Status of Critical Infrastructure as an Object of Cyber Threats," in Laing, C., Badii, A. and Vickers, P. (eds) *Securing Critical Infrastructures and Critical Control Systems: Approaches for Threat Protection*, Hershey, PA: IGI Global, pp. 168–89. doi:10.4018/978-1-4666-2659-1.

———. (2013b) "The U.S. Military's Social Media Civil War: Technology as Antagonism in Discourses of Information-Age Conflict," *Cambridge Review of International Affairs*, 27, 2: 226–45. doi:10.1080/09557571.2012.734787.

———. (2014) *Nonlinear Science and Warfare: Chaos, Complexity, and the U.S. Military in the Information Age*, London: Routledge. doi:10.4324/9780203766446.

———. (2015) "Bringing the 'Cyber' Back Into U.S. Cyberwar Discourse," *Georgetown Journal of International Affairs*, October: 212–22.

Lawson, S. and Gehl, R.W. (2011) *Convergence Security: Cyber-Surveillance and the Biopolitical Production of Security*, Presented at Cyber-Surveillance in Everyday Life: An International Workshop, University of Toronto.

Lawson, S. and Middleton, M.K. (2019) "Cyber Pearl Harbor: Analogy, Fear, and the Framing of Cyber Security Threats in the United States, 1991–2016," *First Monday*, 24, 3. doi:10.5210/fm.v24i3.9623.

Lewis, J.A. (2010) "The Cyber War Has Not Begun," unpublished manuscript.

———. (2017) "The Truth About a Cyber Pearl Harbor," *CNN*, 29 August 2017. Online. Available: <www.cnn.com/2017/08/29/opinions/truth-about-cyber-pearl-harbor-lewis/index.html> (accessed 29 August 2017).

———. (2018) "Cognitive Effect and State Conflict in Cyberspace," unpublished manuscript.

Lipton, E., Sanger, D.E. and Shane, S. (2016) "The Perfect Weapon: How Russian Cyberpower Invaded the U.S.," *New York Times*, 13 December 2016. Online. Available: <www.nytimes.com/2016/12/13/us/politics/russia-hack-election-dnc.html?nytmobile=0&_r=1> (accessed 13 December 2016).

Luhmann, N. (1993) "Deconstruction as Second-Order Observing," *New Literary History*, 24, 4: 763–82. doi:10.2307/469391.

———. (2000) "Why Does Society Describe Itself as Postmodern?," in Rasch, W. and Wolfe, C. (eds) *Observing Complexity: Systems Theory and Postmodernity*, Minneapolis: University of Minnesota Press, pp. 35–50.

Lyngaas, S. (2015) "NSA's Rogers Makes the Case for Cyber Norms," *FCW*, 23 February 2015. Online. Available: <https://fcw.com/articles/2015/02/23/nsa-rogers-cyber-norms.aspx> (accessed 23 February 2015).

Lynn III, W.J. (2010) "Defending a New Domain: The Pentagon's Cyberstrategy," *Foreign Affairs*, 89, 5: 97–108.

———. (2011) *Remarks on the Department of Defense Cyber Strategy*, Presented at National Defense University, 14 July 2011.

Makus, A. (1990) "Stuart Hall's Theory of Ideology: A Frame for Rhetorical Criticism," *Western Journal of Communication*, 54, 4: 495–514. doi:10.1080/10570319009374357.

Marcus, G.E. (1995) *Technoscientific Imaginaries: Conversations, Profiles, and Memoirs*, Chicago: University of Chicago Press. doi:10.1016/0898-1221(95)90189-2.

Markoff, J. (1999) "Blown to Bits; Cyberwarfare Breaks the Rules of Military Engagement," *The New York Times*, 17 October 1999, LexisNexis.

Marks, J. (2019) "The Cybersecurity 202: 2020 Hopeful Seth Moulton Is Calling for a 'Cyber Wall.' Here are the Details," *The Washington Post*, 20 May 2019. Online. Available: <www.washingtonpost.com/news/powerpost/paloma/the-cybersecurity-202/2019/06/20/the-cybersecurity-202-2020-hopeful-seth-moulton-is-calling-for-a-cyber-wall-here-are-the-details/5d0acc51a7a0a47d87c56d9c/?noredirect=on&utm_term=.57c205242061> (accessed 20 May 2019).

Marsh, R.T. (1997) *Critical Foundations: Protecting America's Infrastructures: The Report of the President's Commission on Critical Infrastructure Protection*, Washington, DC: The White House.

Martinez, J. (2012) "Napolitano: US Financial Institutions 'Actively Under Attack' by Hackers," *The Hill*, 31 October 2012. Online. Available: <https://thehill.com/policy/technology/265167-napolitano-us-financial-institutions-qactively-under-attackq-by-hackers> (accessed 31 October 2012).

McConnell, M. (2010) "Mike McConnell on How to Win the Cyber-war We're Losing," *The Washington Post*, 28 February 2010, B01.

McMillan, R. and Valentino-DeVries, J. (2016) "Russian Hacks Show Cybersecurity Limits," *Wall Street Journal*, 1 November 2016. Online. Available: <www.wsj.com/articles/russian-hacks-show-cybersecurity-limits-1478031535> (accessed 1 November 2016).

Mecklin, J. (2019) "A New Abnormal: It Is Still 2 Minutes to Midnight," *Bulletin of the Atomic Scientists*, 24 January 2019. Online. Available: <https://thebulletin.org/doomsday-clock/current-time/> (accessed 24 January 2019).

Meyer, D. (2010) "Cyberwar Could Be Worse Than a Tsunami," *ZDNet*, 3 September 2010. Online. Available: <www.zdnet.com/news/cyberwar-could-be-worse-than-a-tsunami/462576> (accessed 3 September 2010).

Motherboard. (2019) "Why There's No Need to Panic About a 'Cyber 9/11'," *Motherboard*, 14 May 2019. Online. Available: <www.vice.com/en_us/article/ywy3z7/no-need-to-panic-about-cyber-911> (accessed 14 May 2019).

Mueller III, R.S. (2019) *Report on the Investigation Into Russian Interference in the 2016 Presidential Election, Volume I*, Washington, DC: United States Department of Justice.

Muller, B.J. (2008) "Securing the Political Imagination: Popular Culture, the Security Dispositif and the Biometric State," *Security Dialogue*, 39, 2–3: 199–220. doi:10.1177/0967010608088775.

Nexon, D. (2012) "Skyfall and Cyberwar: James Bond Enters the Digital Era," *Duck of Minerva*, 10 November 2012. Online. Available: <https://duckofminerva.com/2012/11/skyfall-and-cyberwar-james-bond-enters-the-digital-era.html> (accessed 10 November 2012).

Nichol, J. (2008) *Russia-Georgia Conflict in South Ossetia: Context and Implications for U.S. Interests*, Washington, DC: Congressional Research Service.

Norton UK. (2012) "Skyfall—the latest 007's enemy is cyberterrorism," *Norton UK*, 16 November 2012. Online. Available : <www://uk.norton.com/skyfall/article> (accessed 16 November 2012).

Nye Jr, J.S. (2016) "The Kremlin and the US election," *Project Syndicate*, 5 December 2016. Online. Available: <www.project-syndicate.org/commentary/kremlin-cyber-attacks-american-election-by-joseph-s--nye-2016-12?barrier=accessreg> (accessed 5 December 2016).

————. (2019) "Deterrence in Cyberspace," *The Strategist*, 7 June 2019. Online. Available: <www.aspistrategist.org.au/deterrence-in-cyberspace/> (accessed 7 June 2019).

Obama, B. (2015) "Remarks by the President at the Cybersecurity and Consumer Protection Summit," *Office of the Press Secretary, The White House*, 13 February 2015. Online. Available: <www.whitehouse.gov/the-press-office/2015/02/13/remarks-president-cybersecurity-and-consumer-protection-summit> (accessed 13 February 2015).

Office of the Director of National Intelligence. (2017) *Assessing Russian Activities and Intentions in Recent U.S. Elections*, Washington, DC: Office of the Director of National Intelligence.

Orcutt, M. (2017) "Sizing Up Trump's Cyberwar Strategy," *MIT Technology Review*, 24 March 2017. Online. Available: <www.technologyreview.com/s/603898/sizing-up-trumps-cyberwar-strategy/> (accessed 24 March 2017).

Ottis, R. (2010) "The Vulnerability of the Information Society," *futureGOV Asia Pacific*, August–September 2010, 70–2.

Pagliery, J. (2015) "Senate Overwhelmingly Passes Historic Cybersecurity Bill," *CNN*, 27 October 2015. Online. Available: <https://money.cnn.com/2015/10/27/technology/cisa-cybersecurity-information-sharing-act/> (accessed 27 October 2015).

Panetta, L. (2012) *Defending the National From Cyber Attacks*, Presented at Business Executives for National Security, New York, NY, 11 October 2012.

Panetta, L. and Talent, J. (2019) "Public-private Joint Effort Is Needed to Prevent a Cyber Pearl Harbor," *The Hill*, 2 April 2019. Online. Available: <https://thehill.com/opinion/cybersecurity/436902-we-need-public-private-collaboration-to-prevent-a-cyber-pearl-harbor> (accessed 2 April 2019).

Patterson, T. (2010) "The Pentagon's Latest Cyber War Games, Told From the Inside," *Federal Computer Week*, 2010. Online. Available: <https://fcw.com/Articles/2010/10/12/Inside-Pentagon-cyber-war-game.aspx> (accessed 2010).

Pierce, S. (2012) "How Real Is 'Skyfall's' Portrayal of Cyberterrorism?," *CNN*, 14 November 2012. Online. Available: <www.cnn.com/2012/11/14/showbiz/movies/skyfall-cyberterrorism-pierce/> (accessed 14 November 2012).

Pollard, N.A. and Devost, M.G. (2016) "Is Cyberwar Turning Out to Be Very Different From What We Thought?," *Politico*, 6 August 2016. Online. Available: <www.politico.com/magazine/story/2016/08/is-cyberwar-turning-out-to-be-very-different-from-what-we-thought-214136> (accessed 6 August 2016).

Pollard, N.A., Segal, A.M. and Devost, M.G. (2018) "Trust War: Dangerous Trends in Cyber Conflict," *War on the Rocks*, 16 January 2018. Online. Available: <https://warontherocks.com/2018/01/trust-war-dangerous-trends-cyber-conflict/> (accessed 16 January 2018).

Pollock, J. (2017) "Russian Disinformation Technology," *MIT Technology Review*, 13 April 2017. Online. Available: <www.technologyreview.com/s/604084/russian-disinformation-technology/> (accessed 13 April 2017).

Porup, J.M. (2019) "Add Cybersecurity to Doomsday Clock Concerns, Says Bulletin of Atomic Scientists," *CSO Online*, 6 February 2019. Online. Available: <www.csoonline.com/article/3338102/add-cybersecurity-to-doomsday-clock-concerns-says-bulletin-of-atomic-scientists.html> (accessed 6 February 2019).

Poulsen, K. (2007) "'Cyberwar' and Estonia's Panic Attack," *Threat Level*, 22 August 2007. Online. Available: <www.wired.com/threatlevel/2007/08/cyber-war-and-e> (accessed 22 August 2007).

Pretorius, J. (2008) "The Security Imaginary: Explaining Military Isomorphism," *Security Dialogue*, 39, 1: 99–120.

Rhodes, R. (2011) *Cyber Meltdown: Bible Prophecy and the Imminent Threat of Cyberterrorism*, Eugene, OR: Harvest House Publishers.

Rid, T. (2011) "Cyber War Will Not Take Place," *Journal of Strategic Studies*, 35, 1: 5–32. doi:10.1080/01402390.2011.608939.

Rid, T. and Buchanan, B. (2018) "Hacking Democracy," *SAIS Review of International Affairs*, 38, 1: 3–16. doi:10.1353/sais.2018.0001.

Rothkopf, D. (2011a) "Where Fukushima Meets Stuxnet: The Growing Threat of Cyber War," *Foreign Policy*, 17 March 2011. Online. Available: <https://foreignpolicy.com/2011/03/17/where-fukushima-meets-stuxnet-the-growing-threat-of-cyber-war/> (accessed 17 March 2011).

———. (2011b) "The Phantom War Has Begun," *Foreign Policy*, 3 November 2011b. Online. Available: <https://foreignpolicy.com/2011/11/03/the-phantom-war-has-begun/> (accessed 3 November 2011b).

Salter, M.B. (2011) "When Securitization Fails: The Hard Case of Counter-Terrorism Programs," in Balzacq, T. (ed) *Securitization Theory: How Security Problems Emerge and Dissolve*, New York: Routledge, pp. 116–31. doi:10.4324/9780203868508.

Sanger, D.E. (2015a) "U.S. Decides to Retaliate Against China's Hacking," *New York Times*, 31 July 2015a. Online. Available: <www.nytimes.com/2015/08/01/world/asia/us-decides-to-retaliate-against-chinas-hacking.html> (accessed 31 July 2015a).

———. (2015b) "Hackers Took Fingerprints of 5.6 Million U.S. Workers, Government Says," *New York Times*, 23 September 2015b. Online. Available: <www.nytimes.com/2015/09/24/world/asia/hackers-took-fingerprints-of-5-6-million-us-workers-government-says.html> (accessed 23 September 2015b).

Sanger, D.E. and Perloth, N. (2014) "U.S. Said to Find North Korea Ordered Cyberattack on Sony," *New York Times*, 17 December 2014. Online. Available: <www.nytimes.com/2014/12/18/world/asia/us-links-north-korea-to-sony-hacking.html> (accessed 17 December 2014).

Schneier, B. (2010) *Keynote Address*, Presented at Conference on Cyber Conflict, Cooperative Cyber Defence Centre of Excellence, Tallinn, Estonia, 18 June 2010.

———. (2014) "Quantum Technology Sold by Cyberweapons Arms Manufacturers," *Schneier on Security*, 14 August 2014. Online. Available: <www.schneier.com/blog/archives/2014/08/quantum_technol.html> (accessed 14 August 2014).

Schulte, S.R. (2013) *Cached: Decoding the Internet in Global Popular Culture*, New York: New York University Press. doi:10.18574/nyu/9780814708668.001.0001.

Schwartau, W. (1991a) *Terminal Compromise*, Seminole, FL: Inter-Pact Press.

———. (1991b) "Fighting Terminal Terrorism," *Computerworld*, 28 January 1991b, 23.

———. (n.d.) "Biography," *Winn Schwartau: Civilian Architect of InfoWar*, n.d. Online. Available: <https://winnschwartau.com/biography/> (accessed 11/14/2015).

Senate Armed Services Committee. (2015a) *Statement for the Record: Worldwide Threat Assessment of the U.S. Intelligence Community*, United States Senate, 26 February 2015a.

Sikka, T. (2008) "Ballistic Missile Defense and Articulation Theory: An Analysis of Technology Using a Cultural Studies Approach," *Journal of Language and Politics*, 7, 1: 119–35. doi:10.1075/jlp.7.1.06sik.

Singel, R. (2009) "Is the Hacking Threat to National Security Overblown?," *Threat Level*, 3 June 2009. Online. Available: <www.wired.com/threatlevel/2009/06/cyberthreat> (accessed 3 June 2009).

Slack, J.D. (2006) "Communication as Articulation," in Shepherd, G.J., St. John, J. and Striphas, T.G. (eds) *Communication as . . : Perspectives on Theory*, Thousand Oaks, CA: Sage Publications. doi:10.4135/9781483329055.

Sommer, P. and Brown, I. (2011) *Reducing Systemic Cybersecurity Risk*, Paris: OECD.

Spring, T. (2017) "Attackers Redefining Objectives, Approaches," *Threat Post*, 5 October 2017. Online. Available: <https://threatpost.com/attackers-redefining-objectives-approaches/128276/> (accessed 5 October 2017).

Stevens, T. (2016) *Power in and Through Cyberspace*, Presentation. International Conference on Cyber Conflict, Tallinn, Estonia. Online. Available: <www.youtube.com/watch?v=Y14ThzI0o5k>.

Stiehm, J. (2002) *The U.S. Army War College: Military Education in a Democracy*, Philadelphia: Temple University Press.

Stohl, M. (2007) "Cyber Terrorism: A Clear and Present Danger, the Sum of All Fears, Breaking Point or Patriot Games?," *Crime, Law and Social Change*, 46, 4–5: 223–38. doi:10.1007/s10611-007-9061-9.

Stone, A. (2019) "How Leon Panetta's 'Cyber Pearl Harbor' Warning Shaped Cyber Command," *Fifth Domain*, 30 July 2019. Online. Available: <www.fifthdomain.com/opinion/2019/07/30/how-leon-panettas-cyber-pearl-harbor-warning-shaped-cyber-command/> (accessed 2019).

Taylor, C. (2004) *Modern Social Imaginaries*, Raleigh: Duke University Press. doi:10.1215/9780822385806.

Thierer, A. (2013) "Technopanics, Threat Inflation, and the Danger of an Information Technology Precautionary Principle," *Minnesota Journal of Law, Science & Technology*, 14: 309. doi:10.2139/ssrn.2012494.

U.S. Senate Select Committee on Intelligence. (2018) *Committee Findings on the 2017 Intelligence Community Assessment*, Washington, DC: U.S. Senate Select Committee on Intelligence.

Valeriano, B. and Jensen, B.M. (2019) "The Myth of the Cyber Offense: The Case for Restraint," *CATO Policy Analysis*, No. 862: 1–16.

Valeriano, B., Jensen, B.M. and Maness, R.C. (2018) *Cyber Strategy: The Evolving Character of Power and Coercion*, New York, NY: Oxford University Press. doi:10.1093/oso/9780190618094.001.0001.

Valeriano, B. and Maness, R.C. (2015) *Cyber War Versus Cyber Realities: Cyber Conflict in the International System*, London: Oxford University Press. doi:10.1093/acprof:oso/9780190204792.001.0001.

Von Drehle, D. (2017) "We're at Cyberwar. And the Enemy Is Us," *The Washington Post*, 17 November 2017. Online. Available: <www.washingtonpost.com/opinions/were-at-cyberwar-and-were-our-own-worst-enemy/2017/11/17/7fb3d522-cbca-11e7-aa96-54417592cf72_story.html?noredirect=on&utm_term=.9af7d5cc258b> (accessed 17 November 2017).

Walt, S.M. (2010) "Is the Cyber Threat Overblown?," *Foreign Policy*, 30 March 2010. Online. Available: <https://walt.foreignpolicy.com/posts/2010/03/30/is_the_cyber_threat_overblown> (accessed 30 March 2010).

Waugh, R. (2012) "Bond's Most Realistic Enemy Ever? Cyber Experts Say Skyfall Risk Is 'Real'," *Yahoo! News*, 7 November 2012. Online. Available: <https://uk.news.yahoo.com/bond-s-most-realistic-enemy-ever-cyber-experts-say-skyfall-risk-is--real-.html#OFgTfew> (accessed 7 November 2012).

Weimann, G. (2005) "Cyberterrorism: The Sum of All Fears?," *Studies in Conflict & Terrorism*, 28, 2: 129–49. doi:10.1080/10576100590905110.

———. (2008) "Cyber-Terrorism: Are We Barking At the Wrong Tree?," *Harvard Asia Pacific Review*, 9, 2: 41–6.

Weinstein, D. (2018) "Stop Saying 'Digital Pearl Harbor'," *Dark Reading*, 2 October 2018. Online. Available: <www.darkreading.com/threat-intelligence/stop-saying-digital-pearl-harbor/a/d-id/1332932> (accessed 2 October 2018).

Weisman, S. (2015) "The Hacking of OPM: Is It Our Cyber 9/11?," *USA Today*, 13 June 2015.Online.Available:<www.usatoday.com/story/money/columnist/2015/06/13/hacking-opm-weisman/28697915/> (accessed 13 June 2015).

Weldes, J. (1996) "Constructing National Interests," *European Journal of International Relations*, 2, 3: 275–318. doi:10.1177/1354066196002003001.

———. (1999) "The Cultural Production of Crises: U.S. Identity and Missiles in Cuba," in Weldes, J., et al. (eds) *Cultures of Insecurity: States, Communities, and the Production of Danger*, Minneapolis, MN: University of Minnesota Press, pp. 35–62.

The White House. (2009a) *Cyberspace Policy Review: Assuring a Resilient Information and Communications Infrastructure*, Washington, DC: The White House.

———. (2009b) "Remarks By the President on Securing Our Nation's Cyber Infrastructure," *White House Press Office*, 29 May 2009b. Online. Available: <www.whitehouse.gov/the_press_office/Remarks-by-the-President-on-Securing-Our-Nations-Cyber-Infrastructure/> (accessed 29 May 2009b).

———. (2011) *International Strategy for Cyberspace: Prosperity, Security, and Openness in a Networked World*, Washington, DC: The White House.

Whitehouse, S. (2011) "Cybersecurity Needs Complete Plan," *Politico*, 7 March 2011. Online. Available: <https://dyn.politico.com/printstory.cfm?uuid=8D587513-F55D-BEB0-A25C03B5B2018326> (accessed 7 March 2011).

Wolfe, J. (2018) "The National Intelligence Director Issued a Warning About a Cyber 9/11-like Cyberattack," *Slate Magazine*, 19 July 2018. Online. Available: <https://slate.com/technology/2018/07/u-s-intel-chief-warns-of-a-crippling-cyberattack-against-our-critical-infrastructure-what-does-he-mean.html> (accessed 19 July 2018).

Yakabuski, M. (2012) "Bond vs. Stuxnet: 4 Ways to Fend Off Cyber Attacks," *SafeNet*, 12 November 2012. Online. Available: <https://data-protection.safenet-inc.com/2012/11/bond-vs-stuxnet-4-ways-to-fend-off-cyber-attacks/> (accessed 12 November 2012).

2 From *WarGames* to cyber Pearl Harbor

Motivating cybersecurity with cyber-doom rhetoric

Introduction

On June 27, 1991, novelist and computer security entrepreneur Winn Schwartau testified before the Subcommittee on Technology and Competitiveness of the U.S. House of Representatives Committee on Science, Space, and Technology. In his testimony, he warned of the potential for a crippling attack against U.S. computer networks that would be akin to what he called an "electronic Pearl Harbor." Though Schwartau is largely credited with being the first to use the Pearl Harbor analogy to warn of the potential impacts of a large-scale cyberattack (Stevens, 2016: 131), his testimony that day was not the first or only time he would use the analogy. He had introduced the phrase "electronic Pearl Harbor" earlier that same year in an article for *Computerworld* magazine titled, "Fighting Terminal Terrorism," as well as in a self-published novel, *Terminal Compromise*, published the same month as his Congressional testimony (Schwartau, 1991a, b). As we will see later in this chapter, the hypothetical cyber Pearl Harbor scenario has been one of the most prominent scenarios used to sell fear of cyber-doom over the last twenty-five years.

As we saw in the first chapter, variations on the phrase "electronic Pearl Harbor" persist in U.S. public policy discourse to this day, as do the fears that this analogy expressed. As such, Schwartau, his novel, article, and testimony are exemplary of the complexity of cyber-doom rhetoric in the American public policy discourse about cybersecurity, which is the focus of this chapter. The fictional "electronic Pearl Harbor" demonstrates the strange mixing of past and future, fact and fiction, in cyber-doom rhetoric. Schwartau himself demonstrates that multiple kinds of speakers use such rhetoric – sometimes in the figure of the same individual – in this case novelist, technology entrepreneur, and expert witness. Such rhetoric appears in a number of genres, including popular fiction, news media, official hearings, and more. Finally, we see how quickly such rhetoric can travel between speakers and genres, in this case from novelist and novel to expert witness and Congressional hearing. Schwartau's case helps to demonstrate that cyber-doom rhetoric makes use of a number of tactics, including historical analogy, in an attempt to call attention to present cybersecurity challenges and promote a response to them. In this rhetoric, past, present, and future blur and, as they do, so do fact and fiction.

The primary goal of this chapter is to begin answering the question of why, after so many years of failed predictions of imminent cyber-doom and even outright rejection by many experts and officials, warnings like Schwartau's "electronic Pearl Harbor" nonetheless persist in U.S. public policy discourse about cyber-security. The first reason, this chapter will argue, is that fear sells, or at least we often believe that it does. We will examine the appeal of fear as a means of raising awareness and motivating audiences to act and, in particular, the rhetorical tactics used to promote fear of cyber-doom in the United States. As Schwartau's case demonstrates, historical analogies and hypothetical scenarios are two of the most prominent of such tactics. This chapter will also identify a number of other tactics used to sell fear of cyber-doom. In short, this chapter argues that one reason for cyber-doom rhetoric's persistence can be found in the appeal of fear appeals and the tactics used to express our fears of cyber-doom.

The appeal of/to fear

Though sometimes more explicit and other times more subtle, nonetheless, cyber-doom rhetoric operates by appealing to fear as a way of calling attention and motivating a response to cybersecurity challenges. Thus, in this section, I will argue that one reason why cyber-doom rhetoric persists is that we have a tendency to believe that appeals to fear are an effective form of persuasion, so much so that cyber-doom rhetoric is exemplary of a wider culture and politics of fear in the United States. Many of the rhetorical tactics associated with the perpetuation of our culture of fear are at work in cyber-doom rhetoric as well.

Various observers have identified the prominence of appeals to fear in the U.S. public policy debate about cybersecurity. Even experts who may otherwise hold opposing views on cyber threats and responses have noted this tendency. Bill Blunden, a computer security expert and harsh critic of U.S. cybersecurity policy, argues, "The core message of cyberwar isn't designed to appeal to reason. It's disseminated in a manner that appeals to emotion" (Blunden and Cheung, 2014: 8763). Similarly, another computer security expert and critic, Marcus Ranum, has argued that cybersecurity "pundits vie with their willing victims in an effort to scare them with ever-worsening nightmares" (Ranum, 2012). But even those who are decidedly more sympathetic to the official U.S. position on cyber threats and responses have noted the tendency towards the use of fear appeals. For example, James A. Lewis, an influential cybersecurity expert at the Center for Strategic and International Studies, has argued, "Some analysts believe that without exaggeration we may never see the United States take this threat seriously. [. . .] There is some merit to these arguments. Appeals to emotions like fear can be more compelling than a rational discussion of strategy" (Lewis, 2010: 4). As mentioned in Chapter 1, several scholars have argued that this is true in the case of cyber-doom scenarios because they tap into and amplify a number of deeper cultural fears and anxieties, including fear of the unknown, new technology, and terrorism (Embar-Seddon, 2002; Stohl and Stohl, 2007; Weimann, 2008).

Research in a number of disciplines, including communication studies, sociology, psychology, and law, lend support to these observations. First, in rare instances, policymakers have admitted to using fear to raise awareness and promote change. For example, as we saw in Chapter 1, former Secretary of Defense Leon Panetta has admitted that his use of a cyber Pearl Harbor doom scenario in 2012 was intended as "shock therapy" meant to "shake up" and "awaken" Americans to the dangers of cyber threats (Stone, 2019). In her cultural history of the Internet, communication scholar Stephanie Schulte examined the influential role that the 1983 movie *WarGames* played in shaping early U.S. computer security laws. In this Hollywood blockbuster, a teenaged hacker breaks into computer systems at NORAD and almost starts a nuclear war. Legislators concerned with the possibility of such a scenario held hearings and even showed clips of the film on the floor of the U.S. Congress. The hope, one lawmaker admitted, was that "the anxiety generated or tapped by" the film, combined with a string of sensationalized media accounts of real-world hacking incidents that were themselves viewed through the lens of the movie *WarGames*, would lead to action on computer security (Schulte, 2013: 21, 26–7, 48–9).

Second, scholars have long noted the appeal of fear appeals as tools for persuasion. Douglas Walton, who has written most extensively on the topic, notes that while social scientists have focused in recent decades on the effectiveness of fear appeals, the tradition of studying fear as a tool of persuasion has deep historical roots in Western scholarship. Rhetoricians, logicians, and scholars of argumentation as far back as Aristotle have studied the logical structure, variations, and ethics of fear appeal arguments. Based on his review of this work, he provides several definitions of fear appeal, which capture the complexity and subtlety of this form of rhetoric. On one hand, a fear appeal can involve "a kind of argument used to threaten a target audience with a fearful outcome (most typically that outcome is the likelihood of death), in order to get the audience to adopt some recommended response" (Walton, 2000: 1). On the other, a fear appeal can involve "a warning that some bad or scary outcome will occur if the respondent does not carry out a recommended action" (Walton, 2000: 1). In still other cases, Walton observes that "the fear appeal argument can be more complex in its structure." In these cases, he explains that fear appeals work based on suggestion, slippery slope, and uncertainty about a possible future:

> What is alleged by the speaker is that if the hearer carries out one action, then that will lead to another, and so forth, in a sequence of connected events that results in some horrible or fearful outcome. What is fearful for the respondent in this type of argument can be not only the final outcome, but also the uncertainty and insecurity attached to the uncontrollability of this sequence. Some dangerous event that, it is said, might happen in the future, raises gloomy foreboding and fears related to the uncontrollability of what could possibly happen in an uncertain world. Fear appeal arguments [of this type] trade on uncertainty about a possible future sequence of events that might be set into motion once a step in a certain direction is taken.
>
> (Walton, 2000: 14–15)

Taken together, these definitions help us to understand why scholars have traditionally taken a dim view of fear appeals, often condemning them as logical fallacies or as unethical. For example, the key difference between the first two definitions expressed earlier is that in the first, the speaker himself is the one threatening the audience with a fearful outcome while, in the second, the speaker is merely warning of a danger. Thus, scholars have tended to condemn all fear appeals because some fear appeals can be based on overt threats (e.g. as in a classic protection racket). Similarly, in the more complex, third definition, the concern is that the user of fear appeal is not necessarily appealing to fear of a legitimate and likely danger, but rather, to the audience's generic fear of an uncertain future, a condition not likely to be alleviated as the future will always remain uncertain.

However, more recent scholarship, including that of Walton, but also rhetorician Michael Pfau (Pfau, 2007), argues that not all fear appeals are fallacious and sets out to distinguish when such tools of persuasion can be ethically permissible. Such a determination cannot be made *a priori*, but rather, within a particular context of communication (Walton, 2000: Ch. 7). Thus, before we can address the question of the ethics and effectiveness of cyber-doom rhetoric (a topic to which we will return in Chapter 5), we must first understand cyber-doom rhetoric as relying on appeal to fear and the various tactics that it employs to do so. In the next section, we will see that cyber-doom appeals to fear of an uncertain and dangerous possible future. But we will also see that some common tactics of cyber-doom rhetoric veer into the territory of argument from threat, the kind that has traditionally worried scholars the most, causing them to label all fear appeals as fallacies.

Regardless of whether they are ethical and effective, the appeal of fear appeals exerts a strong pull. Political scientist Corey Robin has traced the appeal of fear as an idea in Western political thought, demonstrating that

> We – or those who write on our behalf – seem to like the idea of being afraid. Not because fear alerts us to real danger or propels us to take action against it, but because fear is supposed to arouse a heightened state of experience. It quickens our perceptions as no other emotion can, forcing us to see and to act in the world in new and more interesting ways, with greater moral discrimination and a more acute consciousness of our surroundings and ourselves.
>
> (Robin, 2004: 3–4)

Similarly, Walton notes that we have the tendency to believe that if a little fear is good, then the more fear we invoke in audiences, the more effective our arguments will be. We see this, for example, in early social science work that used fear appeals for public health campaigns such as promotion of seat belts or smoking cessation (Walton, 2000: 17). Though the research eventually showed that more fear does not, in fact, lead to more effective messaging (a point we will explore in more detail in Chapter 5), nonetheless, the temptation to ramp up the fear component to get peoples' attention and motivate action remains a strong temptation, especially among some policymakers and activists (Peters et al., 2013, 2014).

This leads us to the third reason why cybersecurity advocates have so often succumbed to the seeming appeal of fear appeals. Though fear appeals have clearly been with us for a long time, several scholars have argued that fear has come to dominate our public discourse and worldview, including our culture, politics, and laws. The culture and politics of fear are marked by a heightened sense of fear on the part of the public and use of fear appeals on the part of politicians, experts, and activists. Legal scholar Cass Sunstein argues that "officials have acted as worst-case entrepreneurs, attempting to ensure that people consider the worst that might happen" in an effort "to energize people" and "to shake people from complacency and spur them into immediate action" (Sunstein, 2007: 272, 180, and 184). Indeed, some political and social theorists, activists, and politicians consciously embrace the promotion of fear and anxiety, even through the use of exaggeration or lies in some cases, as a positive political force that can rally and unite citizens around common fears, especially in a political environment where a positive vision for the future is lacking (Furedi, 2005: 135–9; Robin, 2004).

Sociologists like Barry Glassner, Frank Furedi, and David Altheide have led the way in defining and documenting the emergence of a culture and politics of fear in the West generally, and the United States in particular. Furedi argues, "It is not hope but fear that excites and shapes the cultural imagination of the early twenty-first century" (Furedi, 2006: vii), with the result that

> The politics of fear appears to dominate public life in Western societies. We have become very good at scaring one another and appearing scared. [. . .] From the food we consume to our anxieties about children, being scared has become a culturally sanctioned affectation that pervades all aspects of life. [. . .] [T]he practice has been internalized by the entire political class and has become institutionalized in public life.
>
> (Furedi, 2005: 1)

Though it is common to believe that fear has come to dominate our culture and politics as a result of the terrorist attacks of September 11, 2001, Furedi argues that fear has become "common currency" and "widely assimilated" into public discourse of all kinds, not just terrorism (Furedi, 2005: 130). It is a point supported by the work of fellow sociologists Glassner and Altheide, both of whom noted the emergence of a culture and politics of fear in the United States long before the events of September 11 (Altheide, 1997; Altheide and Michalowski, 1999; Glassner, 1999).

This emergence of a culture and politics of fear is important for our understanding of cyber-doom rhetoric. We are all enmeshed within networks of discourse that shape and constrain the way we observe and speak about the world around us. As Glassner argues, "The success of a scare depends not only on how well it is expressed but also . . . on how well it expresses deeper cultural anxieties" (Glassner, 1999: 208). Those who would appeal to fear are shaped and constrained by the common repertoire of tactics for expressing fear and the available well of cultural anxieties to be harnessed. Furedi makes a similar point when he observes that although the politics of fear involves politicians and other advocates using the power of fear appeals to gain attention, defeat their opponents, and push their agendas, fear is also a larger cultural force over which they lack complete control. That is, though various advocates use and promote fear to manipulate the public, nonetheless, they "are themselves habitually overwhelmed by it" (Furedi, 2005: 123–4). Robin concurs, writing, "Though most modern writers and politicians oppose political fear as the enemy of liberty, reason, and other Enlightenment values, they often embrace it, in spite of themselves, as a source of political vitality" (Robin, 2004: 4).

As we will see in the next section, this has been the case with cyber-doom rhetoric too, which has mirrored the larger cultural and political discourses of fear. The manifestation of the culture and politics of fear in the realm of national security has involved a widened, post-Cold War national security agenda in many Western nations. This agenda has included a host of seemingly ambiguous, uncertain, but dangerous "new threats" related to environmental degradation, poverty, health, immigration, and technology (Furedi, 2009: 25–6; Bigo, 2000; Bigo, 2006; Buzan et al., 1998; Hardt and Negri, 2004). Scholars have identified fear of cyber-doom as exemplary of these "new threats" (Wall, 2008: 866; Dunn Cavelty, 2008: 5; Furedi, 2009: 25–6). Thus, like other worst-case entrepreneurs, those who appeal to the fear of cyber-doom are not entirely in control of the rhetoric that they deploy. We should expect, therefore, that the patterns of thought, speech, and expectation that make up the wider culture and politics of fear in the United States would be reflected in cyber-doom rhetoric. Indeed, as we will see in the next section, this has been the case.

Selling fear of cyber-doom

Those who would use fear of future cyber-doom to raise awareness and motivate a response to present cybersecurity challenges have learned well the tactics for selling fear that have worked so well in other areas of public life. As we saw in the first section, focusing an audience's attention on the vision of an uncertain and potentially dangerous future is a common variation of the fear appeal. Sociologist Barry Glassner quotes the master of suspense, the filmmaker Alfred Hitchcock, as saying, "There is no terror in the bang, only in the anticipation of it." Glassner argues, "Fear managers regularly put [Hitchcock's] wisdom to use by depicting would-be perils as imminent disasters." The future is now, and as present and future blur, he says, the blurring of fact and fiction too becomes one of the key tactics of those who would sell fear (Glassner, 1999: 3). This blurring of fact and fiction, news and entertainment, what Glassner calls the "cuisinart effect" (Glassner, 1999: xxxii), is a key feature of the culture and politics of fear (Furedi, 2005: 130; Furedi, 2006: vii, ix–x; Wall, 2008; Altheide, 2006: 2, 47; Altheide, 1997: 648, 650).

In the remainder of this section, we will explore a number of tactics employed in the rhetoric of cyber-doom. Of course, the most recognized tactics are related to the "cuisinart effect," especially the use of hypothetical scenarios, but also popular and official fictions, as frames for understanding the present state of, and possible futures for, cyber threats. Perhaps the second most common tactic is what we will call appropriation. In its most common form, this involves the use of historical analogies like "cyber Pearl Harbor," which appropriates the fear and anxiety evoked by the memory of this event to warn of the potential for a future of cyber-doom. But, as we will see, appropriation involves more than just historical analogy. Beyond fiction and appropriation, users of cyber-doom rhetoric employ a number of other tactics. These include conflation and strategic use of ambiguity, exaggeration and counterfactuals, determinism, and projection. Each of these tactics, individually or in combination, works in its own way to blur the lines between past, present, and future and, as a result, the line between fact and fiction.

Cuisinart effect

As Cass Sunstein has observed, government officials often serve as "worst-case entrepreneurs," using doom scenarios to raise awareness of, and promote responses to, particular problems (Sunstein, 2007: 272). Glassner says, simply, that for these promoters of fear, "scenarios substitute for facts" (Glassner, 1999: 3). The result, Furedi goes so far as to claim, is that much of our fear today is no longer based on actual experiences but instead focuses on fictional, hypothetical, future threats (Furedi, 2006: ix–x).

Officials do not accomplish this feat on their own, but rather, with substantial assistance from news and entertainment media. This occurs in a number of ways and involves the blurring of boundaries and migration of content between various genres of nonfiction and fiction, both popular and official. Popular nonfiction can include news reporting, documentary film, popular nonfiction books, and academic research. Official nonfiction includes the production of documents like government reports and assessments, statements of policy or strategy, and even legislation. Of course, popular fiction includes products like films and novels. But we can also observe a genre that might be called official fiction, which includes government or its agents producing wargames, simulations, scenarios, and what one scholar has called "fantasy documents" (Clarke, 1999).

The emergence of our culture and politics of fear has benefited greatly from the "mashing together of images and story lines from fiction and reality" (Glassner, 1999: xxxii) resulting in a complex system of feedback and reinforcement in and among the various genres of fact and fiction. News media not only report on the facts of the world as uncovered by their own or government investigations, but also on the products of official fictions like wargames, scenarios, and simulations. Similarly, the format of news reporting itself has evolved in the direction of so-called infotainment that "joins entertainment with reality" (Altheide, 2006: 47). This sometimes involves news or educational media producing televised wargames of their own, or even so-called docudramas, all meant to help viewers imagine frightening possible futures. In still other cases, popular fiction can shape the perceptions of officials, reporting by news media, and, as a result, policy deliberations and decisions (Wall, 2008: 862; Schulte, 2013; Furedi, 2006: vii; Kaplan, 2016). All of these crossings of fact and fiction, official and popular, play a role in cyber-doom rhetoric.

First, current and former officials, government agencies, and private actors of various sorts, sometimes in collaboration – one form of the oft-mentioned "public-private partnership" in cybersecurity – produce a steady stream of cyber threat fictions, including wargames, imagined worst-case scenarios, and simulations. In turn, these become fodder shaping news media coverage of cybersecurity. For example, we began this book with a news media report describing the "real scenarios" involving cyberattack-induced "chaos in our streets" produced in a series of 2010 cyber wargames carried out at the National Cyber Research Park, a public-private partnership (Patterson, 2010). There is no shortage of other such examples going all the way back to the 1990s.

In 1997, for example, Eligible Receiver became one of the most widely reported cyber wargames, gaining not just media attention, but also shaping official U.S. cybersecurity policy for decades to come (Glass, 1998; Kaplan, 2016: Ch4; Martelle, 2018). But it was certainly not the first, nor would it be the only, cyber wargame to make headlines over the years. As early as 1995, *The Washington Post* reported on the results of a Defense Information Systems Agency (DISA) simulation meant to test the defenses of DOD networks. The result, the newspaper reported, was that the head of the agency's information warfare division concluded, "We are not prepared for an electronic version of Pearl Harbor" (Munro, 1995).

News media have reported on several Naval War College games over the years, including one in 1997 with Cantor Fitzgerald on Wall Street that included "information warfare attacks . . . launched against critical infrastructures undergirding the U.S. business and financial community." *The Journal of Commerce* used the game to warn of the "next Pearl Harbor-style sneak attack" (Platt, 1997). On the one-year anniversary of the September 11, 2001, terrorist attacks (9/11 attacks), *Investor's Business Daily* reported on another Naval War College game, conducted that time with Gartner Inc. and "more than 100 industry participants." The game "analyzed what-if attacks against four 'critical infrastructures' of society: telecommunications, the Internet, the electric power grid and financial services." Attendees were reported to have been skeptical at the start about whether cyberterrorism was possible, a view that was ultimately supported by the reported results of the game, which demonstrated that pulling off cyber-doom style attacks was more difficult than many had assumed. Nonetheless, by the end of the game, Gartner Vice President French Caldwell was reported to have said, "The biggest surprise was the result of a survey of attendees. They were asked: 'Do you think a digital Pearl Harbor is possible?' They all said yes" (Howell, 2002).

This was just one of a number of cyber wargames to pop up after the 9/11 attacks, many of which ended up getting media attention of their own, as well as sparking the creation of yet more games and scenarios meant to test officials' and experts' ability to respond to impending cyber-doom (Radcliff, 2002). This has continued into recent years. In 2011, as part of a long story explaining the growing threat of a "cyber arms race," the *Christian Science Monitor* reported on hackers breaking into an industrial plant to cause the spill of toxic chemicals. After a few paragraphs, however, the reader learns that the scenario was not real, but rather, a simulation carried out by the Department of Homeland Security (Clayton, 2011). Similarly, in 2013, the *Christian Science Monitor* opened an article with a laundry list of cyber-doom scenarios involving derailing trains, assassinating politicians, poisoning water supplies, cutting electricity, and more. Like the 2011 story, the reader later learns that these are all fictional scenarios carried out at "CyberCity," described as "one of the U.S. military's premier cyberwar simulators." CyberCity scenarios, we are told, provide a glimpse into what "military officials fear most," which the article said was the kind of cyber Pearl Harbor scenario that former Defense Secretary Leon Panetta warned about in 2012 (Mulrine, 2013). In mid-June 2016, as we were receiving the first reports of Russian hacking of the Democratic National Committee, the trade publication *Military Times* led an article on DOD "preparations for a 'cyber 9/11'" with a harrowing tale of cyberattacks resulting in rolling blackouts leaving millions without power, damaged oil refineries, and disabled ports. It was "total mayhem," the report said. Five paragraphs in, however, the report assured readers that this was just a "fictitious scenario" being used in that year's public-private Cyber Guard exercise. "While Cyber Guard is a classified event," the reader is told, "*Military Times* was part of a small, select group of media granted rare access to the exercise's final day" (Tilghman, 2016).

Other actors too have engaged in the promotion of fictionalized cyber-doom, which then shapes official and popular understandings of cybersecurity. Government contractors of various kinds engage in war gaming and scenario construction at the behest of government. In the mid-1990s, for example, the RAND Corporation's "Day After Tomorrow" study was widely cited in official and popular discourses about cybersecurity (Carlin, 1997; Anderson and Hearn, 1996). RAND even took its cyber wargame show on the road and to the U.S. Congress. In 1996, Senator Sam Nunn (D-GA) opened a hearing on "the potential for cyber-based strategic attack" by stating that "intelligence . . . can only take us so far." Instead, he said, we would need to rely on games and simulations to think about possible futures and our available responses. Thus, he welcomed the "unique opportunity to explore these questions today in the setting of an actual wargame scenario presented by our witnesses from the Rand Corporation. This scenario will hopefully provide the subcommittee and the public at large with a better appreciation for the difficult issues which must be wrestled with when it comes to information warfare" (Permanent Investigations Subcommittee Of The Senate Governmental Affairs Committee, 1996).

Additionally, Winn Schwartau was not the only private individual to get in on the act. For example, the husband and wife futurist duo, Alvin and Heidi Toffler, also warned of the possibility for crippling cyberattacks in their books and in interviews with news media (Elias, 1994). The Tofflers' views on the future of Information-Age warfare were extremely influential in the U.S. military and among some politicians, in particular the Republican Speaker of the House of Representatives, Newt Gingrich (R-GA), during the mid- to late 1990s when concern over such scenarios was gaining increased attention in Washington (Masters, 1994; Ehrenreich, 1995; Bunker, 1995; Lawson, 2014: 128; Gingrich, 2016; Smith, 2019). Even when out of government, former officials continue to play an important role in spinning tales of fictional cyber-doom. This can occur through the production of works of popular nonfiction (e.g. Clarke and Knake's, 2010 book, *Cyber War*), gaining these worst-case entrepreneurs attention by the news media. It can also occur through ongoing interaction with other current and former officials and leaders in the private sector. In 2015, for example, former Chief Technology Officer for the Defense Intelligence Agency Bob Gourley's FedCyber conference featured a "Nightmare Round" moderated by none other than Winn Schwartau, in which "five top threat intelligence experts take 10 minutes to describe their worst case nightmare cyber scenario. The audience will then vote on which scenario scares them the most!"[1]

In other cases, news and educational media outlets go a step further and produce their own depictions of fictionalized cyber-doom with the help of government or industry experts. We see an example of this as early as 1995. After reporting on the results of the DISA exercise mentioned earlier, *The Washington Post* reporter inserted a scenario of his own into the story:

> Apply that judgment to a scenario in 1997: Saddam is in his bunker and his troops are again fighting their way into Kuwait City. U.S. troops are airborne to rescue the Emir again. But what if the Iraqis respond to the U.S. intervention by attacking the New York phone system? Their weapon would not be a Scud missile or a bunch of terrorists but a professional hacker sitting in an Amsterdam apartment or an Ivy League-trained Iraqi computer scientist resting in Finland, either of whom could use the Internet to vandalize New York's phone exchanges. Without a phone network, Wall Street goes silent, the city's cash registers stop ringing, scheduled flights to JFK and LaGuardia are rerouted and Howard Stern and the daytime soaps go off the air. Maybe even the computer-controlled power grid goes down along the East Coast, causing widespread panic, looting and rioting.
>
> (Munro, 1995)

In this case, then, we see an example of news media reporting on the results of an official "fiction," a simulation or exercise, to promote fear of cyber Pearl Harbor, with the news report further amplifying the message by contributing a fictional scenario of its own.

We see a similar situation in 1999 when *Fox News* aired the program *Danger on the Internet Highway: Cyberterror*. Based on simulations conducted by John Arquilla at the Naval Postgraduate School, the fictional scenario depicted in the show – which the narrator assures viewers "is not science fiction!" – is set just one year in the future, in the summer of 2000. This cyber war starts with militants hacking websites and posting a manifesto demanding the removal of all U.S. troops from the Middle East. It rapidly escalates, however, to include cyberterrorists causing aircraft to crash, power to go out, trains to collide, nuclear reactors to meltdown, and the stock market to crash. Throughout the scenario, government is portrayed as relatively helpless and no remedies are offered that the viewer might take to help mitigate the threat (Debrix, 2001: 154–5). As a fear appeal message, the program focused on the threat component of severity while ignoring response efficacy, a pattern that is common in mass media fear appeals (Peters et al., 2013: S9).

There have been at least four other, similar examples in the last decade. In 2010, *CNN* aired portions of a wargame called "Cyber Shockwave." The scenario was designed by former CIA Director Michael Hayden for the Bipartisan Policy Center and in collaboration with various industry and academic partners. A number of former government officials played the roles of key cabinet officials, and they were given the task of wrestling with how to respond to a rapidly unfolding cyberattack. The fictional attack involved the use of a malicious smart phone application to spread malware to tens of millions of people across the United States. Not only did this end up knocking out cell phone service to over 20 million users nationwide, it eventually led to the loss of electricity for 10 million customers on the East Coast and explosions at electric substations in Mississippi and Tennessee. Fearing widespread panic, the former officials playing the game advised the President to take a tough, militarized response. This included recommending he "invoke war-time authorities" and "use his Article II Constitutional powers to nationalize utilities and call out the National Guard" (Bipartisan Policy Center, 2010a: 15).

In October 2012, *The Washington Post* undertook a similar exercise. Noting Defense Secretary Panetta's warning about the possibility of a devastating cyber Pearl Harbor attack, the paper convened a group of experts, including current and former government officials, for a "forum focused on strengthening the cyber defenses of institutions critical to the nation: financial firms, electric power grids, transportation and water systems." As part of this forum, "The Post designed a cyber attack scenario – a fictitious attack on a U.S. oil company that paralyzed the company" and asked the experts to discuss how the government would and should respond. The lessons of the exercise were a familiar refrain: "Private industry and government are going to have to work more closely together." The *Post*'s report on the forum ended by quoting FBI Director Robert Mueller saying, "I am convinced that there are only two types of companies: those that have been hacked and those that will be" (*TheWashington Post*, 2012). As we will see in a later section, this familiar refrain is an example of the use of the tactic of determinism.

In other cases, the news or policymaking value of news media portrayals of cyber-doom scenarios is even thinner. This is the case, for example, in the 2012 *Cybergeddon* web-series created by Yahoo! in conjunction with computer security company Symantec (Goldberg, 2012). Directed by the creator of the hit television series *CSI*, *Cybergeddon* ended up being offered as a full-length film on Netflix. Meant to depict the realities of cyberterrorism, *Cybergeddon* focused on tales of cyber-doom in which an evil hacker mastermind carries out attacks that drain bank accounts in Hong Kong and disrupt computer systems at a Los Angeles area dam, all while framing an FBI agent for the crime. Throughout, critics noted the heavy-handed product placement for Symantec products, such as Norton Antivirus, making the film seem more like a 90-minute advertisement than anything else (Hale, 2012; Wilkins, 2012).

One might have expected better from the *National Geographic Channel*, but such was not the case with its 2013 docudrama, *American Blackout*. Airing in October 2013 to roughly 86 million households in the United States (Baron, 2015), this ninety-minute program depicted the supposed impacts of a cyber-doom scenario. The show begins with two epigraphs meant to lend credibility to the scenario depicted. Like the *Fox News* program mentioned earlier, these epigraphs have the effect of declaring that this fiction is not really fiction. The first, a quote from Dr. Richard Andres of the U.S. National War College, asserts, "[A] massive and well-coordinated cyber attack on the electric grid could devastate the economy and cause a large-scale loss of life." The second explains, "The following program draws upon previous events and expert opinions to imagine what might happen if a catastrophic cyber attack disabled the American power grid." The remainder of the program depicts the expected effects of a ten-day blackout. After only three days, the situation devolves into rioting, looting, and violence resulting in loss of life. Electricity and society are depicted as coterminous, with the loss of the former resulting in the immediate collapse of the latter. Conversely, the show ends abruptly when the electricity is restored, dissolving a tense standoff between a family of doomsday preppers and another group looking for food and water, which is symbolic of the concomitant restoration of a functioning society. Throughout the program, both individuals and government are depicted as largely helpless to respond effectively to the situation. Most response action is depicted as coming from government, whose response is centralized in the federal government and the military. This included using military and riot police, the President declaring a state of emergency, and the federal government taking control of distributing food and water supplies. Even these responses, however, are not depicted as particularly effective. Like the *Fox News* program mentioned earlier, and like most mass media fear appeals, *American Blackout* focused almost entirely on the supposed severity of the scenario with little attention paid to response efficacy. What responses were contemplated echoed the centralized, militarized responses that participants in *CNN*'s *Cyber Shockwave* had recommended.

In other cases, the use of fiction goes beyond reporting on official games and simulations or news media creating its own scenarios to include the use of popular fiction like films or novels. Of course, we began Chapter 1 by noting the serious discussions about cyber threats sparked by the James Bond film *Skyfall* in 2012. And, as we saw at the beginning of this chapter, Winn Schwartau first introduced his idea of an "electronic Pearl Harbor" in a novel. These are not the only examples and, in fact, there is good evidence that popular fiction has influenced not just news media reporting on cybersecurity threats but official policy as well. In fact, as Stephanie Schulte demonstrates, the impact of popular fiction on policymaking is often mediated through the news (Schulte, 2013: 11–13, 28, 47).

As the example of *Skyfall* demonstrates, news media have drawn inspiration from a number of films over the years when covering cybersecurity, from the 1977 spy thriller *Telefon*, to the 2007 installment in the *Die Hard* franchise, *Live Free or Die Hard*. In her work, however, Schulte details the impacts that the 1983 movie, *WarGames*, had at the time on media coverage of computer security issues. In the film, a young hacker played by Matthew Broderick manages to hack into the super computers of the North American Aerospace Defense Command (NORAD). Playing what he thinks is an innocent game with the computer almost results in starting a nuclear war with the Soviets. Schulte documents how news organizations took the film seriously and promoted the idea that it was a realistic scenario. In their ensuing coverage of computer security, news media tended to frame other incidents of computer hacking – no matter how trivial – through the dramatic lens of the film, hyping their impacts and framing computer security generally as a national security issue. This framing, she says, allowed "news media outlets [to] force institutions of military power to answer to fantasy allegations lobbied by a fictional film" (Schulte, 2013: 26).

As a result, *WarGames* offered an early glimpse of how policymakers sometimes "act on or change their political goals under the real, perceived, or cited pressure mobilized by a particular film . . . [or] . . . display symbolic attention, meaning they discuss issues in and through the media" (Schulte, 2013: 189). Specifically, Schulte demonstrates the impact that the film had on members of Congress, who held hearings on computer security, viewed clips of the film, and ultimately passed the Computer Fraud and Abuse Act of 1984 (Schulte, 2013: 21–2). Fred Kaplan adds to the story of *WarGames'* influence by explaining that the writers of the film consulted with a RAND Corporation computer security expert (Kaplan, 2016: 10). What's more, the movie did not just grab the attention of Congress, but President Ronald Reagan too. The President reportedly expressed concern to his advisors and asked them to study what could be done to prevent such a scenario. The result, Kaplan reports, was "a confidential national security decision directive, NSDD-145, signed September 17, 1984, titled 'National Policy on Telecommunications and Automated Information Systems Security.'" This, Kaplan claims, "marked the first time that an American president, or a White House directive, discussed what would come to be called 'cyber warfare'" (Kaplan, 2016: 2). All these years later, we now know that "NSDD-145 placed the National Security Agency in charge of securing all computer servers and networks in the United States" (Kaplan, 2016: 3), another example of how cyber-doom scenarios, when taken seriously, can lead to centralized and militarized policy responses.

But the influence of *WarGames* was not an isolated incident. Popular fiction continued to influence U.S. cybersecurity policymaking at the highest levels during the 1990s. In the wake of the 1995 Oklahoma City bombing, President Bill Clinton created the President's Commission on Critical Infrastructure Protection (PCCIP) to study what could and should be done to protect the vital systems on which the nation relied. Quickly realizing that such systems were not only vulnerable to traditional, kinetic attacks, Commission members searched for an appropriate term to describe the many other avenues of attack on critical infrastructure. Kaplan reports that one of the Commission's members, who had just read William Gibson's 1984 novel, *Neuromancer*, suggested the use of the term "cyberspace," which that novel had introduced to the world. Though other members were initially opposed to the term, it stuck and, as Kaplan explains, "From that point on, the group – and others who studied the issue – would speak of 'cyber crime', 'cyber security', 'cyber war'" (Kaplan, 2016: 45).

During this same time period, the NSA continued its push to exert ever-greater influence over U.S. cybersecurity defenses. In doing so, once again, popular fiction played an important role. Kaplan details, for instance, the impact that the hacker movie, *Sneakers*, had on Director of the NSA Mike McConnell. In fact, Larry Lasker and Walter Parkes, the writers of the film, had written *WarGames* years before. One scene in particular reportedly captured McConnell's attention. The main character tells an old friend,

> The world isn't run by weapons anymore, or energy, or money. It's run by ones and zeroes, little bits of data. It's all just electrons. . . . There's a war out there, old friend, a world war. And it's not about who's got the most bullets. It's about who controls the information: what we see and hear, how we work, what we think. It's all about the information.

McConnell is said to have seen this statement as articulating "the NSA mission statement that he'd been seeking." As a result, he encouraged others at the NSA to watch the film and even held a screening for top officials (Kaplan, 2016: 31–2).

Appropriation

But elaborate, fictional tales of cyber-doom are not always required to capture the attention of news media, policymakers, and publics. In many cases, cybersecurity advocates attempt to appropriate the fear and anxiety elicited by non-cyber events, most commonly instances of war, terrorism, or natural disaster, to raise awareness and promote action. This pattern of making sometimes arbitrary and metaphorical linkages between otherwise unrelated issues, in turn reinforcing fears of both, is not unique to the cybersecurity debate. It is, once again, reflective of a feature of the wider culture and politics of fear discussed earlier (Furedi, 2006: 1, 4–5, 29; Furedi, 2005: 126–8, 140).

Appropriation can be accomplished through the use of analogy and metaphor, or by pointing to particularly dramatic non-cyber contemporary events like conventional terrorist attacks or disasters. We have already seen numerous examples of appropriation via historical analogy and metaphor, with cyber Pearl Harbor serving as perhaps the most common across almost three decades of U.S. cybersecurity discourse (Singer, 2014: 37). It was not long after Winn Schwartau's novel-turned-Congressional testimony warning of a fictional electronic Pearl Harbor that officials picked up this metaphor and scenario and ran with it. Indeed, public officials, more than any other group, have used cyber Pearl Harbor-like doom scenarios to call attention to cybersecurity challenges over the years. When they have indulged in describing and warning about such scenarios, news media were there to amplify their message (Lawson and Middleton, 2019). In some cases, news media promoted the possibility of a cyber Pearl Harbor even when officials themselves remained ambivalent. For example, in 1996, a number of reports implied that then CIA Director John Deutch had warned during Congressional testimony of cyber Pearl Harbor-style attacks. In fact, Director Deutch seemed much less certain about the possibility of such scenarios than his questioner (Messmer, 1996; Weiner, 1996; Smith, 1996; Lawson and Middleton, 2019).

While CIA Director Deutch remained cautious, many other influential policy-makers were raising the alarm about impending cyber-doom. This included key individuals involved in producing the 1997 report of the President's Commission on Critical Infrastructure Protection (PCCIP) that would shape the U.S. cybersecurity debate for years to come. Former Deputy Attorney General Jamie Gorelick and retired General Robert Marsh, for instance, had warned for years about the possibility of a cyber Pearl Harbor scenario (Anthes, 1996; Hamre, 2015). It is unsurprising, therefore, that the final PCCIP report contemplated the possibility of such a scenario. In its first paragraph, the report warned,

> A satchel of dynamite and a truckload of fertilizer and diesel fuel are known terrorist tools. Today, the right command sent over a network to a power generating station's control computer could be just as devastating as a backpack full of explosives, and the perpetrator would be more difficult to identify and apprehend.
>
> (Marsh, 1997: x)

As evidence of why readers should be concerned with such a scenario, the report pointed not to real-world examples of such scenarios, but instead to the 1997 Eligible Receiver wargame and simulation mentioned earlier (Marsh, 1997: 8), but also the 1993 World Trade Center bombing and 1996 Khobar Towers bombing in Saudi Arabia (Marsh, 1997: 5). Of course, the initial impetus for the Commission was not a cyberattack, but rather, the 1995 truck bomb attack in Oklahoma City. Nonetheless, as mentioned earlier, news reporting about the PCCIP report served as another conduit for fear of cyber-doom to make its way into the broader cybersecurity debate and public consciousness.

The Cold War and nuclear weapons have also served as common sources of metaphors and analogies for talking about the scale and danger of cybersecurity threats. Prominent cybersecurity advocates have argued that "[w]e sit at a similar historical moment" to the advent of nuclear weapons and emergence of the Cold War (Clarke, 2009). They have, therefore, called for similarly wide-ranging efforts to address the challenge (Clarke, 2009; McConnell, 2010). Advocates and policymakers have also used the Cold War as a way of describing ongoing cyber conflict between the United States and two of its main adversaries, China and Russia (Griffiths, 2007; Clinton, 2010; Clarke and Knake, 2010: 47–62; Clarke, 2011; Goldman, 2011; Snyder, 2018). The Cold War metaphor and analogy has also invited metaphors and analogies to nuclear weapons and deterrence. Some have gone so far as to claim that cyberattacks may have impacts on par with the use of nuclear weapons while others have called for figuring out how to translate the lessons of nuclear deterrence into cyber deterrence (Chairman of the Joint Chiefs of Staff, 2004: 1; Chairman of the Joint

Chiefs of Staff, 2006: C-1; Kass, 2006: 7; Poulsen, 2007; Libicki, 2009; Harris, 2009; McConnell, 2010).

Conventional terrorist attacks have also served as a common source of metaphors, analogies, and events deployed to direct our attention to cybersecurity threats. Next to cyber Pearl Harbor, cyber 9/11 is the most common metaphor in the U.S. cybersecurity debate (Singer, 2014: 37). In the introduction chapter, we encountered a former DHS cybersecurity official claiming that the U.S. had already experienced a cyber 9/11 but that no one but him had noticed (Yoran, 2010), as well as others describing various cyberattack incidents like the OPM hack and others as a cyber 9/11. Former NSA Director Mike McConnell has even warned that a cyberattack could be more destructive than 9/11 (*The Atlantic*, 2010c). As recently as 2018, Director of National Intelligence Dan Coats said the United States must do more to avoid a cyber 9/11 (Wolfe, 2018). In other cases, however, conventional terrorist attacks are used not as metaphors or analogies for the potential impacts of cyberattacks, but rather, to direct attention towards thinking about cyber threats. We see this in the PCCIP mentioned earlier, where the main events that sparked the formation of the Commission and that were used as justification for taking cyber threats more seriously were all conventional terrorist attacks. More recently, officials seized upon the immediate aftermath of the 2015 ISIS terrorist attacks in Paris to call for various cybersecurity measures (Lawson, 2015a). Once again, news media amplified this framing with the result that, for a time at least, ISIS appeared to be the top cyber threat, surpassing both Russia and China in number of mentions in stories related to cybersecurity (Lawson, 2015b).

Finally, natural and man-made disasters also serve as metaphors, analogies, or events used to describe hypothesized impacts of cyberattacks or to direct our attention to cybersecurity threats. For example, in 2000, Jeffrey Hunker, senior director at the National Security Council's Office of Transnational Threats, said that he worried about an "electronic Exxon Valdez" (Ackerman, 2000). In 2009, Paul Kurtz, a cybersecurity advisor to President Obama's transition team, said he worried about a "cyber Katrina" (Epstein, 2009). The following year, the former head of the United Nations' International Telecommunications Union compared the impacts of prospective cyberattacks to the 2004 Indian Ocean tsunami that killed roughly a quarter million people and caused widespread physical destruction in five countries (Meyer, 2010). In 2011, David Rothkopf, an influential editor of *Foreign Policy* magazine, used the occasion of the Japanese earthquake, tsunami, and resulting Fukushima nuclear disaster to call for greater attention to cybersecurity (Rothkopf, 2011). Similarly, in 2012, former Secretary of Homeland Security Janet Napolitano saw in Superstorm Sandy an opportunity to call attention to "the urgency and the immediacy of the cyber problem; the cyberattacks that we are undergoing and continuing to undergo cannot be overestimated" (Lawson, 2012). Other former officials also got in on the act, as when former CIA Director and Secretary of Defense Leon Panetta renewed his warning about a cyber Pearl Harbor by telling a 2014 cybersecurity symposium organized by Symantec that cyberattacks could cause damage on par with that seen in Superstorm Sandy (Ravindranath, 2014). Electrical blackouts make particularly good fodder for warning of cyber-doom. Some appropriated the 2003 blackout that impacted the U.S. Northeast and parts of Canada to raise concern about an intentional attack on the power grid by terrorists (McCafferty, 2004). Two blackouts in 2019 were appropriated to warn of future cyberattacks. First, though there was no evidence of a malicious cause, a massive blackout that impacted Argentina, Uruguay, and parts of Paraguay for several hours was used to warn of "the risk of cyberwarfare" (Kemp, 2019). Then, a short blackout in the United Kingdom caused some to speculate that it might not be an accident, but rather, "exercising cyberattack squad for the even bigger target (U.S. electricity nets)." Whatever the case, the author of a piece on the incident for *GovTech* wrote, "we can learn a lot from major incidents like this" while we wait for cyber 9/11 or cyber Pearl Harbor, which readers are reminded that "experts have predicted . . . will happen at some point" (Lohrmann, 2019).

Conflation and strategic ambiguity

Behind the use of hypothetical scenarios, metaphors, and analogies, conflation and strategic use of ambiguity are tactics for selling fear of cyber-doom. These involve either intentionally or unintentionally remaining ambiguous in one's definition of key terms, assessment of the nature of the cyber threat – e.g. who threatens what and with what potential impacts – or conflating a number of quite different threats into one, monolithic threat. As Van Evera notes, such "monolith thinking" is a tried and true method for inflating the seeming danger of threats, from the Soviet threat during the Cold War to the threat of a supposed al-Qa'ida-Iraq connection in the immediate wake of 9/11 (Van Evera, 2009: xiv).

Far from weakening cyber-doom's appeal, however, worst-case entrepreneurs may see these tactics as helpful in coping with the fact that most actual cyber threats do not meet the standard criteria for a good fear appeal. Effective fear appeal messages typically contain both threat and efficacy components. This means that the fear appeal communicates that the listener is in danger from a threat, but that there is something efficacious that he or she can do to help prevent or mitigate the threat. More specifically, the threat component of a successful fear appeal will seek to persuade the listener that the threat is both severe in its consequences and that the listener is actually susceptible to the threat. A threat that is likely but not harmful is less likely to be persuasive in motivating a listener to pay attention and to take action. Likewise, a threat that is harmful but unlikely is also less likely to be persuasive. The most effective fear appeals communicate convincingly about both the severity of, and audience susceptibility to, the threat in question, as well as the efficacy of a recommended response (Witte, 1998; Witte and Allen, 2000).

Nonetheless, fear appeals do sometimes rely on the strategic use of ambiguity, especially in cases where one does not know precisely who the target audience of the message will be, or the message is meant to appeal simultaneously to different audiences. Communication researcher Charles Atkin defines strategic ambiguity as a tactic "adapted from the crafty communication practices of corporate executives and political candidates" that involves the use of vagueness in messaging that "allows the individual receivers to draw their own implications based on predispositions; the strategic aspect involves manipulating the message content in a manner that plays off the perceptual tendencies of various subgroups" (Atkin, 2001: 66).

Strategic ambiguity can be used in either the threat or response components of the message. In the case of recommended responses, one might employ strategic ambiguity in an attempt to overcome or bypass audience resistance to the message. That is, ambiguity can be used when an audience does not want to do what the message recommends. Similarly, in the case of the threat component, Atkin says, "Messages can be vague in specifying exactly what is the harmful consequence" (Atkin, 2001: 66). He continues,

> For high-threat messages that seek to emphasize severity of harm, it may be advantageous to cite ambiguous consequences that are not readily observable . . . and thus are not readily refutable by those in a counterarguing mode. Messages might also cite concrete consequences of ambiguous origin . . . for which the audience member can make the attribution that they are due to the risky behavior rather than other sources.
>
> (Atkin, 2001: 66)

Finally, he argues that strategic ambiguity in the use of evidence to support a fear appeal can be used to raise the audience's sense of susceptibility to the threat when actual evidence of such is weak or lacking. He writes,

> It often is important to support persuasive incentives with convincing evidence, particularly to augment the credibility of susceptibility claims. For fear appeals where there is a low level of actual vulnerability, the likelihood of harm can be buttressed by depicting rare but vivid cases rather than underwhelming statistical figures.
>
> (Atkin, 2002: 51)

In short, those who would appeal to fear to get an audience's attention, motivate it to act, or both, would ideally be as clear as possible about the severity and susceptibility of the threat and the efficacy of the recommended response. But in practice, all components might not receive the same (or any) emphasis. In some circumstances, the user of fear appeal might believe it better to retain some ambiguity about exactly what it is that is threatened, by whom or what, with what potential impacts, what should be done in response, and/or some combination of these. In still other cases, no one speaker or group of speakers may be responsible for creating ambiguity, but rather, ambiguity in one or more of the forms mentioned earlier may emerge as a sort of structural feature of the wider discourse.

Such ambiguity has been common in the U.S. public policy debate about cyber threats, likely because most cyber threats do not, individually, offer both severity and susceptibility. On the one hand, cyber threats against personal, business, and government information are widespread today. It seems that hardly a week goes by without another report of a massive corporate or government data breach impacting tens or even hundreds of millions of people. Such threats would seem to rank high in terms of susceptibility. However, in the grand scheme of national security threats, the severity of impacts from these kinds of cyber threats is limited. Despite the appropriation of war, WMD, and natural disaster, or the outright exaggeration of the impacts of cyberattacks as frames for thinking about cyber threats, the fact is that the most widespread cyber threats today do not rise to the same level of severity, thus potentially decreasing their effectiveness as fear appeal.

Conversely, the hypothetical, doom scenarios where critical infrastructures are severely damaged or destroyed, followed by social and economic turmoil, would seem to rank high in terms of severity. However, as noted in the first chapter, many experts agree that these scenarios are not the true face of the cyber threat, with some claiming that they are unlikely and others rejecting them entirely. That is, communicating such scenarios in an effort to raise attention and motivate action on cybersecurity might meet the severity requirement, but likely fails the susceptibility test for an effective fear appeal.

Thus, cyber war proponents have had a tendency to employ strategic ambiguity with regard to defining key terms and identifying exactly who it is that threatens what, with what potential impacts, and with what potential responses, in and through cyberspace. In some cases, officials and experts seem explicitly to embrace the use of ambiguity (Efrony, 2019). For example, Frank Cilluffo, director of the Homeland Security Policy Institute at George Washington University, has suggested, "We want to retain some strategic ambiguity, but at the same time we need to be able to make the case that there are certain attacks that predicate a response." James Lewis suggested that Pentagon officials have adopted this view, saying, "When I talk to people at the Pentagon, I don't find confusion over what's an attack and what's not. But I think they don't want to lay all that out clearly either" (Carroll, 2011). There is evidence to suggest that he is correct in his assessment. Asked what kinds of cyberattacks might provoke a military response from the United States, an unnamed official in the Obama Administration refused to answer, saying instead, "Like most operational things like this, the less said, the better" (Markoff and Shanker, 2010). When the Department of Defense Strategy for Operating in Cyberspace was introduced in July 2011, Deputy Secretary of Defense William Lynn III, who played a leading role in the strategy's development, also refused to define what counts as "cyber war" because he believes that "there is some value in keeping it somewhat ambiguous, as a deterrent" (Carroll, 2011). Thus, it is not merely that cyber threats are ambiguous and uncertain by nature (Dunn Cavelty, 2008; Libicki, 2011, 2012), but also at least partly by design.

In other cases, though less explicit, ambiguity can be observed nonetheless. Some officials and experts have been unwilling or unable to define key terms in the ongoing debate, such as what constitutes "cyber war" and whether it actually exists, or the differences between cyberattacks, probes, and intrusions of various kinds, as well as our susceptibility to those and the severity of their impacts (Lawson, 2010a, b). More subtle still is the emergent ambiguity that comes from the constantly shifting threat perception over time. As discussed in the first chapter, the perception of who threatens what and with what potential impact in and through cyberspace has changed over time, from states to non-states actors, infrastructural to information objects, and back again. As noted in the discussion of appropriation earlier, such shifts can occur quite rapidly, as in the wake of the Paris terrorist attacks where we observed a rapid shift in news media concern, from state actors like China and Russia to the potential for terrorist use of cyberattacks, almost overnight.

Ambiguity is also aided by the tendency to conflate high-severity/low-susceptibility and low-severity/high-susceptibility cyber threats into one, monolithic threat, but also at times to conflate vulnerabilities with threats. Cyber conflict as malicious action in or through cyberspace can include crime, "hacktivism" (i.e. hacking for political activism), terrorist use of the Internet, cyberterrorism, espionage, sabotage, and warfare. Just as crime, war, protest, etc. are quite different activities in the "real world," so too are they different when carried out in/through cyberspace. Yet, as a number of observers have noted over the years, they are often conflated in ways that misrepresent and inflate the cybersecurity challenges that we face (Dunn Cavelty, 2008; Lewis, 2009, 2010; Dunn Cavelty, 2010; Rid, 2013).

In some cases, hacktivism has been conflated explicitly or implicitly with terrorism or war, as when WikiLeaks' activities were called "cyberattacks" or "cyber war" by some journalists and officials as far back as 2010 (Thiessen, 2010; Lawson, 2010c), or when the NSA Director warned that the Anonymous hacktivist group might cyberattack the power grid (Gorman, 2012), or when one lawmaker called for a military "show of force" in response to a series of denial of service attacks on U.S. websites (Zetter, 2009), or when the Department of Defense included "activist groups" as potentially relevant threat actors in its 2015 Department of Defense Cyber Strategy (Defense, 2015: 17, 22). We see crime and warfare conflated or the differences made ambiguous when the NSA Director says, "We're still trying to work our way through distinguishing the difference between criminal hacking and an act of war" (Donohue, 2014). Terrorist group use of the Internet is sometimes conflated with cyberterrorism (Conway, 2002; Stohl, 2007; Weimann, 2005, 2008), as in a 2015 FBI briefing that portrayed social media hacking and "doxing" by an ISIS member as terrorism in and of itself (Chan, 2015). Acts of digital sabotage can be conflated with traditional acts of warfare, as when the North Korean hack of Sony was likened to a cyber Pearl Harbor or outright called an act or war (Lyngaas, 2015; Gingrich, 2014). Finally, espionage is sometimes conflated with terrorism or warfare, as when an official at U.S. Strategic Command wondered when "intensive spying" can be said to "cross the line" into attack (Perera, 2009), or when a cyber war expert asserts that cyber espionage-enabled cases of assassination or political repression meet the Clausewitzian definition of war (Lawson, 2011), and when legislators or commentators call the Chinese hack of the Office of Personnel Management worse than 9/11 or the long-awaited fulfillment of cyber Pearl Harbor predictions (Geraghty, 2015; Weisman, 2015). Gen. Michael Hayden best summed up this tendency towards conflation in 2012 when, in testifying before the Senate Homeland Security Committee, he said, "I should add that cyber, terrorist and criminal threats today all merge in a witches' brew of danger" (Radack, 2012).

In still other cases, prophets of cyber-doom may not identify a particular threat, but instead point to generic vulnerabilities as evidence that cyber-doom is coming. Like the other tactics identified in this chapter, this one has been with us from the beginning. For example, though Winn Schwartau's 1991 piece for *Computerworld* warned of a future of cyberterrorism, implying that the main threat was from non-state actors, it did not identify any particular threat. Instead, it focused on a list of risks and vulnerabilities in current systems that, it assumed, terrorists would inevitably exploit (Schwartau, 1991b). Conflation of vulnerabilities with threats works hand-in-hand with appropriation. When worst-case entrepreneurs cannot point to actual examples of the kind of threat of which they warn, they can point instead to some other kind of attack or catastrophe mixed with discussion of "vulnerabilities." This combination can be used to argue explicitly, or merely to imply, that the simultaneous existence of vulnerabilities and bad actors willing to carry out other kinds of attacks means that cyber-doom is inevitable. In short, many assessments jump from the existence of vulnerabilities to the inevitability of attack, merely assuming the existence of actors possessing both the means and motivation to carry out such attacks (Dunn Cavelty, 2008: 103).

Turning fact into fiction

Of course, there are also plenty of actual cyberattacks that serve as focusing events. But in many of these cases, it is not the actual impacts of these events that worst-case entrepreneurs marshal to raise fear of cyber-doom. Instead of these real-world cyber incidents tempering our views of how possible and likely cyber-doom really is, in some cases the opposite is true. Our expectation of ever-impending cyber-doom can become a lens through which actual cyberattacks are viewed, serving to reinforce our preexisting fears and distorting our understanding of what is really taking place. In these cases, our expectations can cause us to take fact and turn it into fiction. One manifestation of this can be found in examples of gross exaggeration of the effects of actual cyber incidents, another in examples of using actual cyber incidents to focus on what might have happened (but didn't), and even examples of seeing cyberattacks where none occurred.

As we saw in the case of appropriation, worst-case entrepreneurs use analogies and metaphors to non-cyber events to warn about what might happen in the future – e.g. Pearl Harbor, 9/11, WMD, etc. In contrast, exaggeration can involve using these same analogies and metaphors to claim that actual cyberattacks already rise to the level of cyber-doom. Several examples of blatant exaggeration have already been mentioned earlier in the book. These included former Department of Homeland Security official Amit Yoran claiming that a series of data breaches were akin to a cyber 9/11, but that no one had noticed. Others have made similar claims, such as the CEO of a cybersecurity company who claimed in 2007 that cyber crime constituted a digital Pearl Harbor "that's already happened" but "that people don't understand it has happened" (Blitstein, 2007). In 2014, Secretary of Defense Ash Carter applied this same logic to the leak of classified NSA data by contractor Edward Snowden, saying, "We had a cyber Pearl Harbor. His name was Edward Snowden" (Whitlock, 2014). One lawmaker claimed that the data breach at the Office of Personnel Management was worse than 9/11 (Carman, 2015). The Director of the National Security Agency and Commander of U.S. Cyber Command, Admiral Michael Rogers, claimed that the hacking of Sony met the definition of a cyber Pearl Harbor, even though there was no death or physical injury caused in that incident (Lyngaas, 2015). More recently, some have called the 2016 Russian cyber operations against the U.S. presidential election the fulfillment of cyber Pearl Harbor (Hertling and McKew, 2018).

As we have already seen, analogies to WMD, in particular nuclear weapons, are a prominent feature of the U.S. cybersecurity debate. Like 9/11 and Pearl Harbor, WMD can serve as the basis for exaggerating the impacts of actual cyberattacks. We saw this in the example of the Speaker of the Estonian Parliament claiming that the denial of service attacks that hit a number of government and private websites in Estonia in 2007 somehow resembled a nuclear attack (Poulsen, 2007). Needless to say, this was a gross exaggeration. We will return to the specifics of the Estonia case in Chapter 4. As we saw earlier, depictions of cyber-doom scenarios in popular fiction, like the one depicted in the movie *WarGames*, served as a lens through which news media viewed and, in some cases, exaggerated the impacts of hacking incidents (Schulte, 2013: 26–7). Finally, in his 2016 history of cyber warfare, journalist Fred Kaplan asserted that a 2007 proof-of-concept test showing that a cyberattack could destroy an electrical generator meant that cyberattacks "might play a strategic role, too, as instruments of leverage or weapons of mass destruction, not unlike that of nuclear weapons" (Kaplan, 2016: 168).

Even on the economic front, we see a tendency towards exaggerating the effects of past and current cyberattack incidents. In September 2010, for example, Director of the National Security Agency and Commander of U.S. Cyber Command, General Keith Alexander, compared the impact of intellectual property theft via cyber espionage with an invading army sacking and looting a city (House Committee on Armed Services, 2010: 4). Senators Sheldon Whitehouse and Jon Kyl justified their proposed Cyber Security Public Awareness Act of 2011 by arguing that cyberattacks are causing "what could be the largest illicit transfer of wealth in world history" (Whitehouse, 2011a) as well as the "loss of countless American jobs" (Whitehouse, 2011b). Chairman of the House Intelligence Committee, Representative Michael Rogers (R-MI), went so far as to claim in 2013 that cyber espionage posed the biggest threat to U.S. economic security (McVeigh and Rushe, 2013).

In Chapter 4, we will examine in more detail several of the most prominent analogies used to promote cyber-doom and why they actually provide evidence that cyber-doom is unlikely. For now, it will suffice to say we have yet to see a level of physical destruction, loss of life, or strategic impact from cyberattacks that is even remotely comparable to events like 9/11, Pearl Harbor, or the use of WMD. The attacks of 9/11 and Pearl Harbor led to massive death and destruction and, ultimately, multiyear, global wars for the United States, one of which is still ongoing. The development and use of nuclear weapons at the close of one of those wars was so horrifying that those weapons have never been used in combat again, though they did play a central role in a fifty-year superpower standoff that brought the world to the brink of global annihilation on at least one occasion. Perhaps one day cyberattacks will have comparable impacts. But any suggestion that past or current cyberattacks are comparable to 9/11, Pearl Harbor, or the use of WMD is pure exaggeration, not a sober assessment of the real impacts of these attacks.

Likewise, though cyber crime is a real, costly, and growing problem, at times claims about the supposed economic impacts of cyberattacks also seem wildly exaggerated. Several studies over the years have argued that use of poor quality data has resulted in inflated estimates of cyber crime's true economic costs (Levchenko et al., 2011; Anderson et al., 2012; Florencio and Herley, 2013; Romanosky, 2016). Merely considering the context in which such claims were being made is enough to call their veracity into question. These warnings came in the years following the 2008 global financial crisis, which saw millions losing their homes, jobs, and retirements, and where the global financial system stood on the edge of collapse. To claim, in that context, that cyberattacks were the biggest economic threat to the United States or were a major threat to U.S. jobs, or to compare them to the sacking and looting of entire cities, was clearly an exaggeration.

In other cases, cyber-doom proponents might be more honest about the actual effects of a cyber incident, but will, nonetheless, use the event to warn about what could have, but did not actually occur. This is the "it could have been worse," counterfactual story. Cyber-doom proponents take it as given that catastrophe is coming (see the next section on determinism). Thus, when particular incidents fail to live up to the hype of predicted doom, they are reinterpreted or reframed as indicators, warnings, or portents of what is to come. They do not shake our predetermined beliefs about what we are certain is to come, but merely serve to reinforce them (Furedi, 2006: 30).

Again, we can find numerous examples of the "it could have been worse," counterfactual tactic in use. In March 2016, for example, Admiral Michael Rogers pointed to the December 2015, cyberattack-induced power outage in Ukraine to warn of the potential for future "destructive behavior against critical infrastructure in the United States" (Gertz, 2016). At least in Admiral Rogers' case he was using a cyberattack against a power grid – albeit in another country and with limited impact (we will discuss this case in more detail in Chapter 4) – to warn of similar, future attacks in the United States. In other cases, however, we have seen examples where cyberattack incidents that had nothing to do with power outages were used to warn that things could yet be worse (Blunden and Cheung, 2014: 4918). Other major cyber incidents have sparked similar responses, including a series of denial of service attacks during the Russian invasion of Georgia in 2008 and an incident where the Chinese appear to have rerouted Internet traffic

(Blunden and Cheung, 2014: 610, 5216). In 2019, a story in *Wired* reported on a series of scans of the U.S. electrical grid by a group of hackers known as "Triton." This "reconnaissance has a more foreboding edge," the reader is told, because of a 2017 attack by the same group "that could have easily turned destructive or even lethal" (Greenberg, 2019). Likewise, an April 2019 incident in which a "cyber event" caused a minor disruption of an electric utility sparked renewed discussion about the potential for a cyber Pearl Harbor or cyber 9/11, even though we later learned that no customers were impacted in the incident (*Motherboard*, 2019).

This tendency to focus on what might have been is not new. It can be witnessed as far back as 1998 in the case of Solar Sunrise. In this incident, DOD officials were initially convinced that a series of network intrusions might finally mark the onset of a real cyber war. Analysts in the DOD's Information Operations Cell considered that the intrusion might have come from Russia, China, or even Iraq. In the end, however, it turned out to be a pair of teenaged hackers from California. Fred Kaplan reports that instead of wondering why their initial assessments of the incident had been so far off base, most officials ended up seeing in the event a portent of what might be to come. If teenagers could carry out such an intrusion, they conjectured, nation states could and would do much worse in the future (Kaplan, 2016: 74–8).

Finally, the expectation of cyber-doom can not only lead us to exaggerate the impacts of actual cyber incidents or to focus on what might have been instead of what was, it can also sometimes result in seeing instances of cyberattack where none actually occurred. This was the case in 2011 when U.S. news media breathlessly reported that a pump at a municipal water system in Illinois had been hacked and destroyed by Russian cyberattackers (Nakashima, 2011a; Todd, 2011). News media quoted cybersecurity experts who claimed that this incident was an example of the threat of cyberattacks against critical infrastructure that so many had warned about for two decades (Bellovin, 2011). But less than a week later, the Department of Homeland Security and FBI reported that the pump failure was not, in fact, due to a cyberattack at all. The pump had failed for other reasons and the network traffic from a Russian IP address that had led some to believe this was a Russian cyberattack was actually a contractor logging in to the system while traveling overseas (ICS-CERT, 2011; Nakashima, 2011b). But the expectation of cyber-doom is so strong that one cybersecurity expert who had prominently promoted the theory that this was a cyberattack (Weiss, 2011) refused for a time to believe these conclusions, maintaining instead that "something doesn't smell right" and suggesting that "DHS is covering something up" (Krebs, 2011).

Again, this tendency is not new. In 1997, *Wired* magazine reported on the outcome of a recent "Day After Tomorrow" wargame where high-ranking officials contemplated a cyber-doom scenario involving, simultaneously, trains crashing, telecommunications systems collapsing, air traffic control systems failing, power outages, and malfunctioning satellites, combined with more traditional attacks. The article quotes Howard Frank, then director of DARPA's Information Technology Office who had managed one of the "Day After" wargame sessions, talking about a series of West Coast electricity blackouts: "Each time I hear about one of these things, I say to myself, 'OK, it's started!' And when I find out it really didn't, I just think we've bought some additional time. But it *will* start" (Carlin, 1997).

In 2002, we can see an incident in which a cyber wargame primed participants to believe that an actual electrical blackout in India was somehow "a test bed for a cyberattack against the U.S." After interviewing French Caldwell of Gartner, the news report claims that, "at the very least, participants come away with a healthy dose of paranoia" (Radcliff, 2002). Even otherwise credible news media can get caught up in the paranoia, as in the 2016 case where *The Washington Post* reported that the Russians had infiltrated the U.S. electric grid via a malware-infected laptop at a Vermont utility. The paper later corrected and a follow-up was issued that essentially retracted the main claims of the original story, which the hoax-busting website, Snopes.com, rated as "mostly false" (Snopes.com, n.d.; Nakashima, 2017; Eilperin and Entous, 2016). That paranoia was still active in 2018 when many speculated that a series of natural gas explosions in Massachusetts was not only the result of a cyberattack, but perhaps somehow tied to the Stuxnet attack that the U.S. and Israel had carried out on Iran in 2010 (see the last section of this chapter on projection). As in the cases mentioned earlier, the incident turned out not to have been a cyberattack (Franceschi-Bicchierai, 2018). Once again, in 2019, reports of a power outage in New York City led former Chief Technology Officer of the Defense Intelligence Agency Bob Gourley to tweet, "And so began the great cyber war of 2019 . . . #cyberwar #CyberSecurity."[2] Yet again, the outage turned out not to have been caused by a cyberattack (Dobnik and Swenson, 2019).

In short, the expectation that comes from imagining and playing cyber-doom for so long can shape our expectations such that we can see, at least initially, otherwise normal accidents and malfunctions as nefarious cyberattacks when, in fact, they are not. The expectation of cyber-doom is so powerful that some, like the expert mentioned earlier, refuse to believe that certain incidents are not cyberattacks. Even those like Frank who accept that their initial beliefs were wrong may, nonetheless, reframe such incidents as portents of ever-impending doom. Indeed, the very fact of cyber-doom's failure to arrive seems to make us more, rather than less, anxious as we continue to wait for the bang (Wall, 2008).

Determinism

In the preceding examples of turning fact into fiction, we also get a glimpse of a key piece of "rhetorical software" used in promoting fear of cyber-doom. Rhetorical software here refers to those constructions and beliefs that are incredibly influential but also so taken for granted that they are effectively "forgotten, ignored, or, what amounts to the same thing, assumed" (Doyle, 1997: 10). Rhetorical software as "logic" or "underlying presupposition" serves as a sort of operating system enabling the operation of other rhetorical tools (Makus, 1990: 498). In this case, that key piece of rhetorical software is the assumption of cyber-doom's inevitability.

Though, like an operating system, rhetorical software often works in the background and largely unnoticed, sometimes it bubbles to the surface or even becomes the explicit focus of attention. We see an example of this when DARPA's Howard Frank explains that he sees cyberattack in each blackout because of his conviction that cyber war *"will* start" even if it has not already (Carlin, 1997). We see it more recently in the Bob Gourley tween mentioned earlier.

When cyber-doom's inevitability is articulated explicitly, it often takes the "not if, but when" or "when, then" form. These forms have a long history and have been a prominent feature of our larger culture and politics of fear (Furedi, 2006: 135), especially in the years since the terrorist attacks of September 11, 2001. Elmer and Opel argue that in these years, especially in the United States, but also throughout the West, we have seen a subtle shift from worrying about "what if" scenarios to a focus on "when, then" scenarios or, what they call, a focus on the "inevitable future" and what we might do now, in the present, to prevent its inevitability (Elmer and Opel, 2008: 12, 19). They write, "'when, then' logic invokes . . . a shared fatalistic assumption, a sense of futility that collectively moves us to an inevitable future, so as to rhetorically 'gaze backward' in an effort to control the future" (Elmer and Opel, 2008: 21). In the "when, then" form applied to cyber-doom, instead of just pondering the hypothetical "what if" scenario, the speaker argues that cyber-doom in some form is a matter of "when not if" and provides some description of the consequences – i.e. when a massive cyberattack happens, then we will be sorry, or, then it will be too late, or, then I will have been proven correct, or, then we will be forced to take a certain action, etc.

Again, there is no shortage of examples from the U.S. cybersecurity debate over the years. The fictional scenario at the heart of the U.S. military's 2016 Cyber Guard exercise included blackouts, degraded financial systems, oil spills, closed ports, and degraded military command and control systems, what *Military Times* described simply as, "total mayhem." U.S. Cyber Command's director of training, Coast Guard Rear Admiral Kevin Lunday, was quoted as saying, "For us, it's not a question of if it will happen but when. The more relevant question is: When it does [happen], will we as a Department of Defense, will we as a nation and with our allies, be ready for it?" (Tilghman, 2016). Admiral Lundy was no doubt taking a cue from his boss, Admiral Rogers, who, speaking to the audience of the RSA security conference in March 2016 about the cyberattack-induced blackouts in Ukraine the prior December, said, "It is only a matter of the 'when,' not the 'if' we're going to see a nation-state, group or actor engage in destructive behavior against critical infrastructure in the United States" (Gertz, 2016).

Once again, the rhetoric of an inevitable future is not new in the U.S. cybersecurity debate. The 1997 PCCIP report mentioned earlier implied that a cyber Pearl Harbor-like scenario was only a matter of time and exhorted, "Waiting for disaster is a dangerous strategy. Now is the time to act to protect our future" (Marsh, 1997: 6). Such sentiment is perhaps to be expected given that one of the earliest and most influential articles on the subject is titled simply and definitively, "Cyberwar is Coming!" (Arquilla and Ronfeldt, 1997). Since that time, one of its authors, John Arquilla, has warned that cyberattack presents "a grave and growing capacity for crippling our tech-dependent society." Without immediate action, he said,

> The alternative will be, inevitably, a cyber 9/11 that could have dire consequences for the economy or for our troops in the field if they are engaged in battle when the digital storm hits. If such an attack does come, no commission will be able to conclude that it could not have been foreseen. The portents have been there for all to see. There can be no excuse for failure to take action now.
>
> (Arquilla, 2009)

As we will see in the next chapter, this combination of assumptions about technological dependence and determinism taps into deep-seated cultural fears and anxieties in the United States, which is part of why cyber-doom fear appeals have had such staying power. But we have also gotten a glimpse of how assumptions of an inevitable future can potentially lead us astray by encouraging exaggeration, counterfactuals, and misdiagnosis to divert our thinking about cybersecurity from the realm of fact to the realm of fiction. This is a topic to which we will return in Chapter 5.

Projection

Another prominent tactic of cyber-doom rhetoric in the U.S. cybersecurity debate is the use of projection. In psychology, projection involves attributing to others those beliefs, desires, or actions that we ourselves harbor or have undertaken but do not wish to acknowledge. Glassner has identified projection at work in our broader culture and politics of fear. In that case, we project or shift blame onto others – which can include individuals, groups, technology, or more nebulous social forces or conditions – problems that we have created ourselves, either actively or through inaction or neglect. As in the case of individual psychology, we sometimes engage in projection as a way of ignoring or simplifying complex, difficult problems that we otherwise have not solved or wish to avoid because they are caused by our own bad policies. In short, he says, "bad guys substitute for bad policies" when it comes time to assign blame (Glassner, 1999: xxxix, 6–8).

In the case of cybersecurity discourse, projection involves pointing to specific malicious incidents, activities, or capabilities contemplated, carried out, or caused by the United States as a way of raising fear of what others might do to the United States in/through cyberspace. After years of being warned about the malicious intent of such adversaries as the Chinese; the Russians; the Iranians; and even unspecified, generic hackers, criminals, and terrorists, we have learned in recent years that the United States is not entirely innocent of some of the very behavior it has projected onto others. In fact, such projection has played a foundational role in U.S. cyber war policy from the beginning.

In his 2016 history of U.S. cyber war policy, Fred Kaplan notes that as early as the late 1970s, U.S. defense officials began to worry that "what the United States was doing to its enemies, its enemies could also do to the United States – maybe not right now, but someday soon" (Kaplan, 2016: 11). Similarly, he reports, "a disturbing thought smacked a few analysts inside NSA: Anything we're doing to them, they can do to us" (Kaplan, 2016: 17). Indeed, as Kaplan's account makes clear, time and again, U.S. concern over the need to improve its cyber defenses was almost entirely rooted in fear of its own desires, capabilities, or actions projected onto others. This was true even when those others had yet to carry out similar actions, were years or decades away from having similar capabilities, and may not have even expressed a similar desire to acquire or use the kinds of capabilities that the United States was already acquiring and using. In short, our fear of malicious cyber others was, from the start, often a fear of ourselves.

Projection has been a surprisingly recurrent theme and motivating factor among some of the highest ranking, most influential voices in the U.S. cyber war debate. In the 1990s, NSA Director Mike McConnell was motivated, in part, by concern, as Kaplan said, that "whatever we can do to our enemies, our enemies could soon do to us" (Kaplan, 2016: 172). Richard Clarke, who would become the first White House "cybersecurity czar" and who remains an influential voice in the cybersecurity debate, came to a similar realization a few years later. Kaplan says, "If we can do this to other countries, he [Clarke] realized, they'll soon be able to do the same thing to us – and that meant we were screwed" (Kaplan, 2016: 90). Finally, after hearing briefings on the threats that cyberattacks posed to the United States, as well as the ongoing offensive cyber campaign by the United States against Iran – i.e. Olympic Games, later dubbed "Stuxnet" in the media – President Obama learned "the obverse of the usual lesson: what the enemy might someday do to us, we can now do to the enemy" (Kaplan, 2016: 202). That is, fear that an adversary might do to us what we desired to do, and for which we were building capabilities, became a justification for actually carrying out preemptive, offensive cyberattacks.

In yet one more twist, however, some officials and experts have used Stuxnet, both directly and indirectly, as evidence of the cyber threat facing the United States. In August 2011, former CIA counterterrorism official Cofer Black warned the audience at the Black Hat hacker conference that the United States was not taking the threat of cyberattack seriously enough, just as it had failed to appreciate the threat posed by terrorism in the years leading up to the attacks of September 11, 2001. In addition to pointing to recent reports at that time of a massive campaign of cyber espionage supposedly carried out by the Chinese, Black also identified Stuxnet as "the Rubicon of our future" (Zakaria, 2011). Like Cofer Black, in the wake of Stuxnet, former CIA and NSA Director General Michael Hayden said, "someone crossed the Rubicon" (Sanger, 2012). Of course, we now know that this signal cyber event that served to focus so much attention on the supposed threat to the United States was itself perpetrated by the United States (Sanger, 2012). This revelation even resulted in criticism from some prominent cybersecurity proponents who are otherwise sympathetic to the cyber war message. Jason Healey, at the time the director of the Cyber Statecraft Initiative at the Atlantic Council, wrote,

The message to the US private sector therefore seems to be that they need to be regulated because they are not protecting themselves sufficiently against a weapon designed and launched by their own government. The arsonist wants to legislate better fire codes.

(Healey, 2012)

Members of Congress have also at times used Stuxnet as evidence of the threat facing the United States. In 2011, Senators Joseph Lieberman (D/I-CT), Susan Collins, and Tom Carper (D-DE) penned an op-ed that used Stuxnet to encourage passage of their proposed cybersecurity legislation, which would allow us to "avoid a cyber Pearl Harbor" (Lieberman et al., 2011). The following year, in a report to Congress, the Government Accountability Office used Stuxnet as evidence of the cyber threat posed to the U.S. electricity grid (Wilshusen and Trimble, 2012).

Despite the concern of folks like Healey, however, Michael Gross best summed up the view of many officials and experts, all while drawing on nuclear and World War II metaphors in the process, an example of how cyber-doom tactics can work in combination. He wrote, "Stuxnet is the Hiroshima of cyber-war. That is its true significance, and all the speculation about its target and its source should not blind us to that larger reality. We have crossed a threshold, and there is no turning back" (Gross, 2011).

That is, pay no attention to the fact this threat is one of our own making. Also do not pay attention to the target of our preemptive cyberattack because, if you do, you will see that yet another set of malicious cyber incidents used as evidence of threat were, in fact, the direct result of our own actions. For example, we now know that the Iranian cyberattacks on Saudi Aramco and several U.S. banks that Secretary Panetta used as evidence of a cyber Pearl Harbor threat to the United States were Iran's reprisal for the Stuxnet and Flame attacks (Kaplan, 2016: 213). Beyond Stuxnet and the reprisals it set off, we also see examples of U.S. officials pointing to security flaws and vulnerabilities as evidence of the threats we face when we know the United States has, in some cases, either worked to create those flaws or allowed them to go unpatched in the belief that they could be exploited to the United States' benefit at a later date (Ball et al., 2013; Gallagher and Greenwald, 2014; Pagliery, 2015; Schneier, 2014; Kaplan, 2016: 17, 34, 92).

Another example helps to demonstrate that projection, like the other tactics of cyber-doom rhetoric discussed earlier, sometimes crosses or combines with other tactics, in this case the use of fiction. Projection can be seen in President Reagan's, his advisors', and Congress' concern over a *WarGames*-like scenario in the early 1980s. This fictional scenario was in part based on reports of poor computer security practices by NORAD officers (Kaplan, 2016: 10) and echoed a number of real-life incidents from the late 1970s and 1980s. But these did not involve an outside, teenage, or foreign hacker. Instead, in the real incidents, human or technical errors led to false alerts at NORAD that, for a time, led U.S. officials to believe that the United States might be under Soviet nuclear attack (Burr, 2012). But instead of focusing policymaker and media attention on what could be done to prevent such unforced errors, a fictional Hollywood blockbuster resulted in a spate of news reports, hearings, an executive order, and legislation all meant to deal with computer security threats posed by nefarious, outsider others.

The tendency towards projection has continued up to the current period. For example, in 2016, as we were only just beginning to learn of Russian cyber operations against the election, an article in the *Sacramento Bee* used Stuxnet as evidence that the possibility of a cyber 9/11 or Pearl Harbor was real (Johnson, 2016). Prior to that, in May of 2016, Defense Secretary Ash Carter pointed to U.S. cyberattacks against ISIS – which officials have called akin to dropping "cyber bombs" (Clark, 2016) – and warned that other countries had similar capabilities that could be used against the United States. "That is why good, strong cyber defenses are essential for us," he said (Menn, 2016). Likewise, a bombshell 2019 report of U.S. hacking of the Russian electrical grid sparked concern about the potential vulnerability of the U.S. grid to such attacks (Sanger and Perloth, 2019; Kemp, 2019).

Conclusion

In this chapter, we examined the first of two reasons for why and how cyber-doom rhetoric persists in U.S. public policy discourse despite years of failed predictions of impending catastrophe and growing recognition that it is not reflective of the vast majority of actual threats. That is, cyber-doom rhetoric works primarily as a tool of persuasion, in particular as an appeal to fear meant to call attention and motivate action. Research in communication, psychology, and sociology has demonstrated that policy makers and advocates find fear appeals to be a particularly appealing form of communication, believing that scaring audiences is the most effective way to get their attention and motivate them. This belief has become so common in recent decades that several prominent sociologists have argued that our culture and politics have come to be dominated by fear, not just of cyber-doom, but of crime, disease, natural disaster, environmental degradation, technology, terrorism, and more. Like all of us, cybersecurity advocates are enmeshed within larger systems of rhetoric, discourse, and culture that shape and constrain the available resources for making their claims. In a society increasingly dominated by fear of all kinds, and with a long tradition of viewing fear as a powerful tool of persuasion, it is perhaps unsurprising that cybersecurity advocates would reach for cyber-doom when making their case. What's more, as we saw earlier, when they do, they choose from a menu of tactics commonly employed by other worst-case entrepreneurs. These include various mixings of fact and fiction; appropriation of the fear and anxiety elicited by other events like war, natural disaster, or terrorism; conflating very different kinds of cyber threats into one monolithic threat; being strategically ambiguous in the use of terms or identification of subjects, objects, and impacts of cyber threats; exaggerating the effects of real incidents or focusing on counterfactuals; portraying cyber-doom as a predetermined reality; and pointing to our own desires, capabilities, or actions as evidence of the threat posed to us by nefarious others. In the aggregate, then, we can say that cyber-doom rhetoric operates as a fear appeal argument that employs a number of rhetorical tactics to reinforce the twin, master metaphors of war and disaster to frame our thinking about the future of cyber conflict, always seemingly calling our attention away from what is to instead focus on the inevitability of future doom.

In the next chapter, we will examine the second reason why cyber-doom rhetoric persists. We have already begun to hint at this reason in this chapter, which is that it is reflective of deeper cultural trends and anxieties. The next chapter argues that some of these trends and anxieties are deeply rooted in American history going all the way back to the nation's founding. In Chapter 4, we will return to some of the key metaphors, analogies, and incidents encountered earlier, demonstrating that they do not, under closer examination, support the case for cyber-doom's inevitability. In Chapter 5, we will explore some of the potentially negative implications of relying on fear appeals generally, war/disaster framing more specifically, as well as identifying determinism and projection in particular as dangerous cognitive distortions.

Notes

1 The program for the "FedCyber 2015 Annual Summit" is available online at http://events.fedcyber.com/ (last accessed 11 June 2019).
2 The URL for the original tweet is https://twitter.com/bobgourley/status/115019783 5993886726. It is also archived at https://archive.fo/MfPuD.

References

Ackerman, E. (2000) "What's Being Done to Make b-to-b Secure," *San Jose Mercury News*, 10 July 2000, 1C.
Altheide, D.L. (1997) "The News Media, the Problem Frame, and the Production of Fear," *Sociological Quarterly*, 38, 4: 647–68. doi:10.1111/j.1533-8525.1997.tb00758.x.
———. (2006) *Terrorism and the Politics of Fear*, Lanham, MD: AltaMira Press.
Altheide, D.L. and Michalowski, R.S. (1999) "Fear in the News: A Discourse of Control," *The Sociological Quarterly*, 40, 3: 475–503. doi:10.1111/j.1533-8525.1999.tb01730.x.
Anderson, R.H. and Hearn, A.C. (1996) *An Exploration of Cyberspace Security R&D Investment Strategies for DARPA: "the Day After.in Cyberspace II"*, Santa Monica, CA: RAND.
Anderson, R.H., et al. (2012) *Measuring the Cost of Cybercrime*, Presentation at 11th Workshop on the Economics of Information Security, Berlin, Germany.
Anthes, G.H. (1996) "White House Launches Cybershield," *Computerworld*, 22 July 1996, 29.
Arquilla, J. (2009) "Click, Click . . . Counting Down to Cyber 9/11," *San Francisco Chronicle*, 26 July 2009, E2.
Arquilla, J. and Ronfeldt, D. (1997) "Cyberwar Is Coming!," in Arquilla, J. and Ronfeldt, D. (eds) *In Athena's Camp: Preparing for Conflict in the Information Age*, Santa Monica, CA: RAND, pp. 24–60.
Atkin, C.K. (2001) "Theory and Principles of Media Health Campaigns," in Rice, R.E. and Atkin, C.K. (eds) *Public Communication Campaigns*, Thousand Oaks, CA: Sage Publications, pp. 49–68. doi:10.4135/9781544308449.

———. (2002) "Promising Strategies for Media Health Campaigns," in Crano, W.D. and Burgoon, M. (eds) *Mass Media and Drug Prevention: Classic and Contemporary Theories and Research*, Mahwah, NJ: L. Erlbaum, pp. 35–66. doi:10.4324/9781410603845.

The Atlantic. (2010c) "Fmr. Intelligence Director: New Cyberattack May Be Worse Than 9/11," *The Atlantic*, 30 September 2010c. Online. Available: <www.theatlantic.com/politics/archive/2010/09/fmr-intelligence-director-new-cyberattack-may-be-worse-than-9-11/63849/> (accessed 30 September 2010c).

Ball, J., Borger, J. and Greenwald, G. (2013) "Revealed: How US and UK Spy Agencies Defeat Internet Privacy and Security," *The Guardian*, 6 September 2013. Online. Available: <www.theguardian.com/world/2013/sep/05/nsa-gchq-encryption-codes-security> (accessed 6 September 2013).

Baron, S. (2015) "List of How Many Homes Each Cable Network Is in as of February 2015," *TV By the Numbers*, 22 February 2015. Online. Available: <https://tvbythenumbers.zap2it.com/reference/list-of-how-many-homes-each-cable-network-is-in-as-of-february-2015/> (accessed 22 February 2015).

Bellovin, S. (2011) "Water Supply System Apparently Hacked, With Physical Damage," *CircleID*, 18 November 2011. Online. Available: <www.circleid.com/posts/20111118_water_supply_system_apparently_hacked_with_physical_damage> (accessed 18 November 2011).

Bigo, D. (2000) "When Two Become One: Internal and External Securitisations in Europe," in Kelstrup, M. and Williams, M. (eds) *International Relations Theory and the Politics of European Integration: Power, Security, and Community*, London: Routledge, pp. 171–204. doi:10.4324/9780203187807.

———. (2006) "Security, Exception, Ban and Surveillance," in Lyon, D. (ed) *Theorizing Surveillance: The Panopticon and Beyond*, Cullompton, Devon: Willan Publishing, pp. 46–58. doi:10.4324/9781843926818.

Bipartisan Policy Center. (2010a) *Cyber Shockwave: Simulation Report and Findings*, Washington, DC: Bipartisan Policy Center.

Blitstein, R. (2007) "Part I: How Online Crooks Put Us All at Risk," *San Jose Mercury News*, 8 November 2007, LexisNexis.

Blunden, W. and Cheung, V. (2014) *Behold a Pale Farce: Cyberwar, Threat Inflation, & the Malware-Industrial Complex*, Waterville, OR: Trine Day.

Bunker, R.J. (1995) "The Tofflerian Paradox," *Military Review*, May–June: 99–102.

Burr, W. (2012) "The 3 a.m. Phone Call," *The Nuclear Vault, The National Security Archive*, 1 March 2012. Online. Available: <https://nsarchive.gwu.edu/nukevault/ebb371/> (accessed 1 March 2012).

Buzan, B., Wæver, O. and Wilde, J.D. (1998) *Security: A New Framework for Analysis*, Boulder, CO: Lynne Rienner Pub.

Carlin, J. (1997) "A Farewell to Arms," *Wired*, 1 May 1997. Online. Available: <www.wired.com/1997/05/netizen-2/> (accessed 1 May 1997).

Carman, A. (2015) "OPM Breaches More Serious to National Security Than 9/11, Congresswoman Argues During Hearing," *SC Magazine*, 16 June 2015. Online. Available: <www.scmagazine.com/house-committee-on-oversight-and-government-reform-hosts-hearing-on-data-breaches/article/421052/> (accessed 16 June 2015).

Carroll, C. (2011) "Congress Demands Cyber Details While DOD Aims for Ambiguity," *Stars and Stripes*, 21 July 2011. Online. Available: <www.stripes.com/news/congress-demands-cyber-details-while-dod-aims-for-ambiguity-1.149790> (accessed 21 July 2011).

Chairman of the Joint Chiefs of Staff. (2004) *The National Military Strategy of the United States of America: A Strategy for Today; a Vision for Tomorrow*, Washington, DC: Chairman of the Joint Chiefs of Staff.

———. (2006) *The National Military Strategy for Cyberspace Operations*, Washington, DC: Chairman of the Joint Chiefs of Staff.

Chan, E. (2015) *Current Threat to the U.S. From Cyber Espionage & Cyberterrorism*, San Francisco, CA: FBI Cyber Division.

Clark, C. (2016) "'It Sucks to Be ISIL:' US Deploys 'Cyber Bombs,' Says Depsecdef," *Breaking Defense*, 12 April 2016. Online. Available: <https://breakingdefense.com/2016/04/it-sucks-to-be-isil-us-deploys-cyber-bombs-says-depsecdef/> (accessed 12 April 2016).

Clarke, L. (1999) *Mission Improbable: Using Fantasy Documents to Tame Disasters*, Chicago: University of Chicago Press.

Clarke, R.A. (2009) "War From Cyberspace," *The National Interest*, October/November.

———. (2011) "China's Cyberassault on America," *Wall Street Journal*, 15 June 2011. Online. Available: <https://online.wsj.com/article/SB100014240527023042593045763733911018288876.html> (accessed 15 June 2011).

Clarke, R.A. and Knake, R. (2010) *Cyber War: The Next Threat to National Security and What to Do About It*, New York: HarperCollins.

Clayton, M. (2011) "The New Cyber Arms Race," *Christian Science Monitor*, 7 March 2011, LexisNexis.

Clinton, H.R. (2010) *Remarks on Internet Freedom*, Presented at The Newseum, Washington, DC, 21 January 2010.

Conway, M. (2002) "Reality Bytes: Cyberterrorism and Terrorist 'Use' of the Internet," *First Monday*, 7, 11. doi:10.5210/fm.v7i11.1001.

Debrix, F. (2001) "Cyberterror and Media-Induced Fears: The Production of Emergency Culture," *Strategies*, 14, 1: 149–68. doi:10.1080/10402130120042415.

Department of Defense. (2015) *Dod Cyber Strategy*, Washington, DC: Department of Defense.

Dobnik, V. and Swenson, A. (2019) "No Lights, Big City: Power Outage KOs Broadway, Times Square," *Associated Press*, 14 July 2019. Online. Available: <https://apnews.com/21bae47cefc24f4bb286384df47dc0bf> (accessed 14 July 2019).

Donohue, B. (2014) "NSA Director Rogers Urges Cyber-Resiliency," *Threat Post*, 16 September 2014. Online. Available: <https://threatpost.com/nsa-director-rogers-urges-cyber-resiliency/108292> (accessed 16 September 2014).

Doyle, R. (1997) *On Beyond Living: Rhetorical Transformations of the Life Sciences*, Stanford, CA: Stanford University Press.

Dunn Cavelty, M. (2008) *Cyber-Security and Threat Politics: U.S. Efforts to Secure the Information Age*, New York: Routledge. doi:10.4324/9780203937419.

———. (2010) "The Real Cyberwar Is About Beating the Crooks and the Spooks," *Parliamentary Brief Online*, 29 October 2010. Online. Available: <www.parliamentarybrief.com/2010/10/the-real-cyberwar-is-about-beating-the-crooks-and-the#all> (accessed 29 October 2010).

Efrony, D. (2019) "Entering the Third Decade of Cyber Threats: Toward Greater Clarity in Cyberspace," *Lawfare Blog*, 13 June 2019. Online. Available: <www.lawfareblog.com/entering-third-decade-cyber-threats-toward-greater-clarity-cyberspace> (accessed 13 June 2019).

Ehrenreich, B. (1995) "Surfing the Third Wave," *The New York Times*, 7 May 1995. Online. Available: <www.nytimes.com/1995/05/07/books/surfing-the-third-wave.html> (accessed 7 May 1995).

Eilperin, J. and Entous, A. (2016) "Russian Operation Hacked a Vermont Utility, Showing Risk to U.S. Electrical Grid Security, Officials Say," *The Washington Post*, 31 December 2016. Online. Available: <www.washingtonpost.com/world/national-security/russian-

hackers-penetrated-us-electricity-grid-through-a-utility-in-vermont/2016/12/30/
8fc90cc4-ceec-11e6-b8a2-8c2a61b0436f_story.html?utm_term=.6ac117d16b0d>
(accessed 31 December 2016).

Elias, T.D. (1994) "Toffler: Computer Attacks Wave of Future," *South Bend Tribune (Indiana)*, 2 January 1994, F10.

Elmer, G. and Opel, A. (2008) *Preempting Dissent: The Politics of an Inevitable Future*, Winnipeg: Arbeiter Ring Publisher.

Embar-Seddon, A. (2002) "Cyberterrorism: Are We Under Siege?," *American Behavioral Scientist*, 45, 6: 1033–43. doi:10.1177/0002764202045006007.

Epstein, K. (2009) "Fearing 'Cyber Katrina,' Obama Candidate for Cyber Czar Urges a 'FEMA for the Internet'," *Business Week*, 18 February 2009. Online. Available: <www.businessweek.com/the_thread/techbeat/archives/2009/02/fearing_cyber_katrina_obama_candidate_for_cyber_czar_urges_a_fema_for_the_internet.html> (accessed 18 February 2009).

Florencio, D. and Herley, C. (2013) "Sex, Lies and Cyber-Crime Surveys," in Schneier, B. (ed) *Economics of Information Security and Privacy III*, New York: Springer, pp. 35–53. doi:10.1007/978-1-4614-1981-5.

Franceschi-Bicchierai, L. (2018) "People Are Recklessly Speculating that the Massachusetts Gas Explosions Were a Stuxnet-related Hack," *Motherboard*, 14 September 2018. Online. Available: <www.vice.com/en_us/article/9kvy5z/massachusetts-gas-explosion-cyberattack-hackers-speculation> (accessed 14 September 2018).

Furedi, F. (2005) *Politics of Fear: Beyond Left and Right*, London: Continuum.

———. (2006) *Culture of Fear Revisited: Risk-Taking and the Morality of Low Expectation*, London: Continuum. doi:10.5040/9781474212427.

———. (2009) *Invitation to Terror: The Expanding Empire of the Unknown*, London: Continuum.

Gallagher, R. and Greenwald, G. (2014) "How the NSA Plans to Infect 'Millions' of Computers With Malware," *The Intercept*, 12 March 2014. Online. Available: <https://firstlook.org/theintercept/2014/03/12/nsa-plans-infect-millions-computers-malware/> (accessed 12 March 2014).

Geraghty, J. (2015) "The OPM Hack Was Just the Start and It Won't Be the Last," *National Review*, 12 June 2015. Online. Available: <www.nationalreview.com/corner/419678/opm-hack-was-just-start-and-it-wont-be-last-jim-geraghty> (accessed 12 June 2015).

Gertz, B. (2016) "Cybercom Says Cyberattacks on Infrastructure Coming," *Washington Times*, 9 March 2016. Online. Available: <www.washingtontimes.com/news/2016/mar/9/inside-the-ring-infrastructure-cyberattacks/> (accessed 9 March 2016).

Gingrich, N. (2014) "America Lost the Cyberwar Over Sony: Now What?," *CNN*, 18 December 2014. Online. Available: <www.cnn.com/2014/12/18/opinion/gingrich-america-lost-cyberwar-sony/> (accessed 18 December 2014).

———. (2016) "Remembering Alvin Toffler," *Politico*, 31 December 2016. Online. Available: <www.politico.com/magazine/story/2016/12/alvin-toffler-obituary-future-shock-214561> (accessed 31 December 2016).

Glass, A.J. (1998) "Target America: Computer Warfare," *The Atlanta Journal and Constitution*, 2 August 1998, 03E.

Glassner, B. (1999) *The Culture of Fear: Why Americans Are Afraid of the Wrong Things*, New York, NY: Basic Books.

Goldberg, L. (2012) "'CSI' Creator Anthony Zuiker Sets 'Cybergeddon' for Yahoo," *The Hollywood Reporter*, 20 March 2012. Online. Available: <www.hollywoodreporter.com/live-feed/csi-anthony-zuiker-yahoo-cybergeddon-302452> (accessed 20 March 2012).

Goldman, D. (2011) "China vs. U.S.: The Cyber Cold War Is Raging," *CNN Money*, 28 July 2011. Online. Available: <https://money.cnn.com/2011/07/28/technology/government_hackers/index.htm> (accessed 28 July 2011).

Gorman, S. (2012) "Alert on Hacker Power Play," *Wall Street Journal*, 21 February 2012. Online. Available: <www.wsj.com/articles/SB1000142405297020405980457722939010 5521090> (accessed 21 February 2012).

Greenberg, A. (2019) "The Highly Dangerous 'Triton' Hackers Have Probed the US Grid," *Wired*, 14 June 2019. Online. Available: <www.wired.com/story/triton-hackers-scan-us-power-grid/> (accessed 14 June 2019).

Griffiths, P. (2007) "World Faces 'Cyber Cold War' Threat," *Reuters*, 29 November 2007. Online. Available: <www.reuters.com/article/2007/11/29/us-britain-internet-idUSL2932083320071129> (accessed 29 November 2007).

Gross, M.J. (2011) "A Declaration of Cyber-war," *Vanity Fair*, 1 April 2011. Online. Available: <www.vanityfair.com/news/2011/04/stuxnet-201104> (accessed 1 April 2011).

Hale, M. (2012) "A Yahoo Series About Cyberterrorism," *The New York Times*, 24 September 2012. Online. Available: <www.nytimes.com/2012/09/25/arts/television/cybergeddon. html> (accessed 24 September 2012).

Hamre, J. (2015) "The 'Electronic Pearl Harbor'," *Politico*, 9 December 2015. Online. Available: <www.politico.com/agenda/story/2015/12/pearl-harbor-cyber-security-war-000335> (accessed 9 December 2015).

Hardt, M. and Negri, A. (2004) *Multitude: War and Democracy in the Age of Empire*, New York: The Penguin Press.

Harris, S. (2009) "The Cyberwar Plan," *National Journal*, 14 November 2009. Online. Available: <www.nationaljournal.com/njmagazine/print_friendly.php?ID=cs_20091114_3145> (accessed 14 November 2009).

Healey, J. (2012) "Stuxnets Are Not in the Us National Interest: An Arsonist Calling for Better Fire Codes," *The New Atlanticist*, 4 June 2012. Online. Available: <www.atlanticcouncil. org/blogs/natosource/stuxnets-are-not-in-the-us-national-interest-an-arsonist-calling-for-better-fire-codes-1> (accessed 4 June 2012).

Hertling, M. and McKew, M.K. (2018) "Putin's Attack on the U.S. Is Our Pearl Harbor," *Politico*, 16 July 2018. Online. Available: <www.politico.com/magazine/story/2018/07/16/putin-russia-trump-2016-pearl-harbor-219015> (accessed 16 July 2018).

House Committee on Armed Services. (2010) *Statement of General Keith B. Alexander Commander United States Cyber Command*, United States House of Representatives, 23 September 2010b.

Howell, D. (2002) "Cyber Terror Lurks as Potential Threat; Could Do Serious Harm; Financial Services and Utilities Among Sectors that Could Be Damaged," *Investor's Business Daily*, 11 September 2002, LexisNexis.

ICS-CERT. (2011) "ICSB-11-327-01-illinois Water Pump Failure Report," *ICS-CERT Information Bulletin*, 23 November 2011. Online. Available: <www.us-cert.gov/control_systems/pdf/ICSB-11-327-01.pdf> (accessed 23 November 2011).

Johnson, T. (2016) "If the NSA Can Be Hacked, Is Anything Safe?," *Sacramento Bee*, 22 August 2016. Online. Available: <www.sacbee.com/news/politics-government/article96471952.html> (accessed 22 August 2016).

Kaplan, F. (2016) *Dark Territory: The Secret History of Cyber War*, New York: Simon & Schuster.

Kass, L. (2006) *Cyberspace: A Warfighting Domain*, Presented at Air Force Cyberspace Task Force, 26 September 2006.

Kemp, J. (2019) "Massive Blackouts and the Risk of Cyberwarfare," *Reuters*, 18 June 2019. Online. Available: <https://uk.reuters.com/article/uk-cyber-electricity-kemp/column-massive-blackouts-and-the-risk-of-cyberwarfare-idUKKCN1TJ1N8> (accessed 18 June 2019).

Krebs, B. (2011) "Dhs Blasts Reports of Illinois Water Station Hack," *Krebs on Security*, 22 November 2011. Online. Available: <https://krebsonsecurity.com/2011/11/dhs-blasts-reports-of-illinois-water-station-hack/> (accessed 22 November 2011).

Lawson, S. (2010a) "Cyberwar: We Don't Know What It Is or If We're in One, But . . .," *Forbes.com*, 23 April 2010a. Online. Available: <www.forbes.com/sites/firewall/2010/04/23/cyberwar-we-dont-know-what-it-is-or-if-were-in-one-but/> (accessed 23 April 2010a).

———. (2010b) "Just How Big Is the Cyber Threat to the Department of Defense?," *Forbes.com*, 4 June 2010b. Online. Available: <www.forbes.com/sites/firewall/2010/06/04/just-how-big-is-the-cyber-threat-to-dod/> (accessed 4 June 2010b).

———. (2010c) "Wikileaks and the Ongoing Beat of the Cyberwar Drums," *Forbes.com*, 9 August 2010c. Online. Available: <www.forbes.com/sites/firewall/2010/08/09/wikileaks-and-the-ongoing-beat-of-the-cyberwar-drums/> (accessed 9 August 2010c).

———. (2011) "Cyber War and the Expanding Definition of War," *Forbes.com*, 26 October 2011. Online. Available: <www.forbes.com/sites/seanlawson/2011/10/26/cyber-war-and-the-expanding-definition-of-war/> (accessed 26 October 2011).

———. (2012) "DHS Secretary Napolitano Uses Hurricane Sandy to Hype Cyber Threat," *Forbes.com*, 1 November 2012. Online. Available: <www.forbes.com/sites/seanlawson/2012/11/01/dhs-secretary-napolitano-uses-hurricane-sandy-to-hype-cyber-threat/> (accessed 1 November 2012).

———. (2014) *Nonlinear Science and Warfare: Chaos, Complexity, and the U.S. Military in the Information Age*, London: Routledge. doi:10.4324/9780203766446.

———. (2015a) "Officials Seize on Paris Attacks to Push Cybersecurity Measures," *Forbes.com*, 5 December 2015a. Online. Available: <www.forbes.com/sites/seanlawson/2015/12/05/officials-seize-on-paris-attacks-to-push-cybersecurity-measures/#787fe2fabf86> (accessed 5 December 2015a).

———. (2015b) "Has ISIS Become the Top Cyber Threat?," *Forbes.com*, 18 December 2015b. Online. Available: <www.forbes.com/sites/seanlawson/2015/12/18/has-isis-become-the-top-cyber-threat/#2b4fdb03159a> (accessed 18 December 2015b).

Lawson, S. and Middleton, M.K. (2019) "Cyber Pearl Harbor: Analogy, Fear, and the Framing of Cyber Security Threats in the United States, 1991–2016," *First Monday*, 24, 3. doi:10.5210/fm.v24i3.9623.

Levchenko, K., et al. (2011) *Click Trajectories: End-to-end Analysis of the Spam Value Chain*, Presentation at Security and Privacy (SP), 2011 IEEE Symposium on Security and Privacy, IEEE. doi:10.1109/sp.2011.24.

Lewis, J.A. (2009) "The Fog of Cyberwar: Discouraging Deterrence," *International Relations and Security Network Special Reports*, 2009. Online. Available: <www.isn.ethz.ch/isn/Current-Affairs/Special-Reports/The-Fog-of-Cyberwar/Deterrence/> (accessed 3 March 2009).

———. (2010) "The Cyber War Has Not Begun," unpublished manuscript.

Libicki, M.C. (2009) *Cyberdeterrence and Cyberwar*, Santa Monica, CA: RAND.

———. (2011) "The Strategic Uses of Ambiguity in Cyberspace," *Military and Strategic Affairs*, 3, 3: 3–10.

———. (2012) "The Specter of Non-Obvious Warfare," *Strategic Studies Quarterly*, 6, 3: 88–101.

Lieberman, S.J., Collins, S.S. and Carper, S.T. (2011) "Avoiding a Cyber Pearl Harbor," *The Washington Post*, 8 July 2011, A13.

Lohrmann, D. (2019) "Lessons From the Massive United Kingdom Power Outage," *GovTech*, 11 August 2019. Online. Available: <www.govtech.com/blogs/lohrmann-on-cybersecurity/lessons-from-the-massive-uk-power-outage.html> (accessed 11 August 2019).

Lyngaas, S. (2015) "NSA's Rogers Makes the Case for Cyber Norms," *FCW*, 23 February 2015. Online. Available: <https://fcw.com/articles/2015/02/23/nsa-rogers-cyber-norms.aspx> (accessed 23 February 2015).

Makus, A. (1990) "Stuart Hall's Theory of Ideology: A Frame for Rhetorical Criticism," *Western Journal of Communication*, 54, 4: 495–514. doi:10.1080/10570319009374357.

Markoff, J. and Shanker, T. (2010) "In Digital Combat, U.S. Finds No Easy Deterrent," *The New York Times*, 25 January 2010. Online. Available: <www.nytimes.com/2010/01/26/world/26cyber.html> (accessed 25 January 2010).

Marsh, R.T. (1997) *Critical Foundations: Protecting America's Infrastructures: The Report of the President's Commission on Critical Infrastructure Protection*, Washington, DC: The White House.

Martelle, M. (2018) "Eligible Receiver 97: Seminal DOD Cyber Exercise Included Mock Terror Strikes and Hostage Simulations," *National Security Archive*, 1 August 2018. Online. Available: <https://nsarchive.gwu.edu/briefing-book/cyber-vault/2018-08-01/eligible-receiver-97-seminal-dod-cyber-exercise-included-mock-terror-strikes-hostage-simulations> (accessed 1 August 2018).

Masters, K. (1994) "Inside Newt's Brain," *The Washington Post*, 12 December 1994. Online. Available: <www.washingtonpost.com/archive/lifestyle/1994/12/12/inside-newts-brain/c49037f9-b66f-4ceb-a6ec-e9c256a0e27c/?utm_term=.3dcae167be73> (accessed 12 December 1994).

McCafferty, D. (2004) "Dark Lessons: Learning From the Blackout of August '03," *Homeland Security Today*, 1 August 2004. Online. Available: <www.hstoday.us/content/view/1177/60/> (accessed 1 August 2004).

McConnell, M. (2010) "Mike McConnell on How to Win the Cyber-war We're Losing," *The Washington Post*, 28 February 2010, B01.

McVeigh, K. and Rushe, D. (2013) "House Passes Cispa Cybersecurity Bill Despite Warnings From White House," *The Guardian*, 18 April 2013. Online. Available: <www.theguardian.com/technology/2013/apr/18/house-representatives-cispa-cybersecurity-white-house-warning> (accessed 18 April 2013).

Menn, J. (2016) "Cyber Attacks on Islamic State Use Tools Others Also Have: U.S. Defense Chief," *Reuters*, 11 May 2016. Online. Available: <www.reuters.com/article/us-cyber-defense-isis/cyber-attacks-on-islamic-state-use-tools-others-also-have-u-s-defense-chief-idUSKCN0Y302F> (accessed 11 May 2016).

Messmer, E. (1996) "CIA Director Calls for Cyber-war Defense Center," *Network World*, 1 July 1996, 8.

Meyer, D. (2010) "Cyberwar Could Be Worse Than a Tsunami," *ZDNet*, 3 September 2010. Online. Available: <www.zdnet.com/news/cyberwar-could-be-worse-than-a-tsunami/462576> (accessed 3 September 2010).

Motherboard. (2019) "Why There's No Need to Panic About a 'Cyber 9/11'," *Motherboard*, 14 May 2019. Online. Available: <www.vice.com/en_us/article/ywy3z7/no-need-to-panic-about-cyber-911> (accessed 14 May 2019).

Mulrine, A. (2013) "Cyber Security: The New Arms Race for a New Front Line," *Christian Science Monitor*, 13 September 2013, LexisNexis.

Munro, N. (1995) "The Pentagon's New Nightmare: An Electronic Pearl Harbor," *The Washington Post*, 16 July 1995, C03.

Nakashima, E. (2011a) "Foreign Hackers Targeted U.S. Water Plant in Apparent Malicious Cyber Attack, Expert Says," *The Washington Post*, 18 November 2011a. Online. Available: <www.washingtonpost.com/blogs/checkpoint-washington/post/foreign-hackers-broke-into-illinois-water-plant-control-system-industry-expert-says/2011/11/18/gIQAgmTZYN_blog.html> (accessed 18 November 2011a).

———. (2011b) "Water-pump Failure in Illinois Wasn't Cyberattack After All," *The Washington Post*, 25 November 2011b. Online. Available: <www.washingtonpost.com/world/national-security/water-pump-failure-in-illinois-wasnt-cyberattack-after-all/2011/11/25/gIQACgTewN_story.html> (accessed 25 November 2011b).

———. (2017) "Russian Government Hackers Do Not Appear to Have Targeted Vermont Utility, Say People Close to Investigation," *The Washington Post*, 2 January 2017. Online. Available: <www.washingtonpost.com/world/national-security/russian-government-hackers-do-not-appear-to-have-targeted-vermont-utility-say-people-close-to-investigation/2017/01/02/70c25956-d12c-11e6-945a-76f69a399dd5_story.html?noredirect=on&utm_term=.c87d9a99e79d> (accessed 2 January 2017).

Pagliery, J. and Perez, E. (2015) "Super-sneaky Malware Found in Companies Worldwide," *CNN.com*, 17 February 2015. Online. Available: <https://money.cnn.com/2015/02/17/technology/security/malware-nsa/> (accessed 17 February 2015).

Patterson, T. (2010) "The Pentagon's Latest Cyber War Games, Told From the Inside," *Federal Computer Week*, 2010. Online. Available: <https://fcw.com/Articles/2010/10/12/Inside-Pentagon-cyber-war-game.aspx> (accessed 2010).

Perera, D. (2009) "Cyber Deterrence Dialog Raises Many Questions," *Defense Systems*, 19 May 2009. Online. Available: <https://defensesystems.com/Articles/2009/05/19/Cyber-deterrence-raises-questions.aspx> (accessed 19 May 2009).

Permanent Investigations Subcommittee of the Senate Governmental Affairs Committee. (1996) *Vulnerability of United States Government Information Systems to Computer Attacks*, United States Senate, 25 June 1996.

Peters, G.-J.Y., Ruiter, R.A.C. and Kok, G. (2013) "Threatening Communication: A Critical Re-Analysis and a Revised Meta-Analytic Test of Fear Appeal Theory," *Health Psychology Review*, 7, sup1: S8–S31. doi:10.1080/17437199.2012.703527.

———. (2014) "Threatening Communication: A Qualitative Study of Fear Appeal Effectiveness Beliefs Among Intervention Developers, Policymakers, Politicians, Scientists, and Advertising Professionals," *International Journal of Psychology*, 49, 2: 71–9. doi:10.1002/ijop.12000.

Pfau, M. (2007) "Who's Afraid of Fear Appeals? Contingency, Courage, and Deliberation in Rhetorical Theory and Practice," *Philosophy and Rhetoric*, 40, 2: 216–37. doi:10.1353/par.2007.0024.

Platt, G. (1997) "New From the Navy: Wall Street War Games," *Journal of Commerce*, 23 December 1997, 1A.

Poulsen, K. (2007) "'Cyberwar' and Estonia's Panic Attack," *Threat Level*, 22 August 2007. Online. Available: <www.wired.com/threatlevel/2007/08/cyber-war-and-e> (accessed 22 August 2007).

Radack, J. (2012) "Hayden Whips Up "Witches' Brew" of Fear in Congress," *DailyKos*, 13 July 2012. Online. Available: <www.dailykos.com/stories/2012/07/13/1109445/-Hayden-Whips-Up-Witches-Brew-of-Fear-in-Congress> (accessed 13 July 2012).

Radcliff, D. (2002) "More Than a Game; Corporations Are Adding Cyberattack Exercises to Their Disaster-preparedness Tactics," *Computerworld*, 9 September 2002, 44.

Ranum, M.J. (2012) *Cyberwar, the Power of Nightmares*, 31 August 2012. Online. Available: <https://fabiusmaximus.com/2012/08/31/42830/> (accessed 4 April 2016).

Ravindranath, M. (2014) "Panetta Warns of 'Cyber Pearl Harbor'," *The Washington Post*, 17 March 2014, A09.

Rid, T. (2013) *Cyber War Will Not Take Place*, Oxford: Oxford University Press.

Robin, C. (2004) *Fear: The History of a Political Idea*, Oxford: Oxford University Press.

Romanosky, S. (2016) "Examining the Costs and Causes of Cyber Incidents," *Journal of Cyber Security*, 2, 2: 121–35. doi:10.1093/cybsec/tyw001.

Rothkopf, D. (2011) "Where Fukushima Meets Stuxnet: The Growing Threat of Cyber War," *Foreign Policy*, 17 March 2011. Online. Available: <https://foreignpolicy. com/2011/03/17/where-fukushima-meets-stuxnet-the-growing-threat-of-cyber-war/> (accessed 17 March 2011).

Sanger, D.E. (2012) "Obama Order Sped Up War of Cyberattacks Against Iran," *New York Times*, 1 June 2012. Online. Available: <www.nytimes.com/2012/06/01/world/middleeast/ obama-ordered-wave-of-cyberattacks-against-iran.html> (accessed 1 June 2012).

Sanger, D.E. and Perloth, N. (2019) "U.S. Escalates Online Attacks on Russia's Power Grid," *The New York Times*, 15 June 2019. Online. Available: <www.nytimes. com/2019/06/15/us/politics/trump-cyber-russia-grid.html> (accessed 15 June 2019).

Schneier, B. (2014) "Quantum Technology Sold By Cyberweapons Arms Manufacturers," *Schneier on Security*, 14 August 2014. Online. Available: <www.schneier.com/blog/ archives/2014/08/quantum_technol.html> (accessed 14 August 2014).

Schulte, S.R. (2013) *Cached: Decoding the Internet in Global Popular Culture*, New York: New York University Press. doi:10.18574/nyu/9780814708668.001.0001.

Schwartau, W. (1991a) *Terminal Compromise*, Seminole, FL: Inter-Pact Press.

———. (1991b) "Fighting Terminal Terrorism," *Computerworld*, 28 January 1991b, 23.

Singer, P.W. (2014) *Cybersecurity and Cyberwar: What Everyone Needs to Know*, London: Oxford University Press.

Smith, H. (2019) "Heidi Toffler, Unheralded Half of a Futurist Writing Team, Dies at 89," *The Washington Post*, 13 February 2019. Online. Available: <www.washingtonpost.com/ local/obituaries/heidi-toffler-unheralded-half-of-a-futurist-writing-team-dies-at-89/2019/02/13/63458c50-2fa6-11e9-8ad3-9a5b113ecd3c_story.html?utm_ term=.6bad0d0b77c1> (accessed 13 February 2019).

Smith, J. (1996) "CIA Gears Up to Thwart 'Information Attacks'," *The Washington Post*, 26 June 1996, A19.

Snopes.com. (n.d.) "Was a Vermont Power Grid Infiltrated by Russian Hackers?," *Snopes. com*, n.d. Online. Available: <www.snopes.com/fact-check/report-vermont-power-grid- infiltrated-by-russian-hackers/> (accessed n.d.).

Snyder, T. (2018) *The Road to Unfreedom: Russia, Europe, America*, New York: Tim Duggan Books.

Stevens, T. (2016) *Power in and Through Cyberspace*, Presentation at International Conference on Cyber Conflict, Tallinn, Estonia. Online. Available: <www.youtube.com/ watch?v=Y14ThzI0o5k>.

Stohl, M. (2007) "Cyber Terrorism: A Clear and Present Danger, the Sum of All Fears, Breaking Point or Patriot Games?," *Crime, Law and Social Change*, 46, 4–5: 223–38. doi:10.1007/s10611-007-9061-9.

Stohl, M. and Stohl, C. (2007) "Networks of Terror: Theoretical Assumptions and Pragmatic Consequences," *Communication Theory*, 17, 2: 93–124. doi:10.1111/j.1468-2885. 2007.00289.x.

Stone, A. (2019) "How Leon Panetta's 'Cyber Pearl Harbor' Warning Shaped Cyber Command," *Fifth Domain*, 30 July 2019. Online. Available: <www.fifthdomain.com/opinion/ 2019/07/30/how-leon-panettas-cyber-pearl-harbor-warning-shaped-cyber- command/> (accessed 30 July 2019).

Subcommittee on Oversight and Investigations, Committee on Energy and Commerce. (2012) *CYBERSECURITY: Challenges in Securing the Modernized Electricity Grid, Statement of Gregory C. Wilshusen, Director Information Security Issues, and David C. Trimble, Director Natural Resources and Environment*, U.S. House of Representatives, 28 February 2012.

Sunstein, C.R. (2007) *Worst-Case Scenarios*, Cambridge, MA: Harvard University Press. doi:10.4159/9780674033535.

Thiessen, M.A. (2010) "Wikileaks Must Be Stopped," *The Washington Post*, 3 August 2010. Online. Available: <www.washingtonpost.com/wp-dyn/content/article/2010/08/02/AR2010080202627_pf.html> (accessed 3 August 2010).

Tilghman, A. (2016) "Inside the Pentagon's Secretive Preparations for a 'Cyber 9/11'," *Military Times*, 21 June 2016. Online. Available: <www.militarytimes.com/story/military/2016/06/21/us-military-cyber-attack-china-russia/86147512/> (accessed 21 June 2016).

Todd, B. (2011) "Water Pumping Station Hacked?," *The Situation Room With Wolf Blitzer, CNN*, 18 November 2011. Online. Available: <https://situationroom.blogs.cnn.com/2011/11/18/water-pumping-station-hacked/> (accessed 18 November 2011).

Van Evera, S. (2009) "Forward," in Thrall, A.T. and Cramer, J.K. (eds) *American Foreign Policy and the Politics of Fear: Threat Inflation Since 9/11*, London: Routledge, pp. xi–xvi. doi:10.4324/9780203879092.

Wall, D.S. (2008) "Cybercrime and the Culture of Fear: Social Science Fiction(s) and the Production of Knowledge About Cybercrime," *Information, Communication & Society*, 11, 6: 861–84. doi:10.1080/13691180802007788.

Walton, D.N. (2000) *Scare Tactics: Arguments That Appeal to Fear and Threats*, Boston: Kluwer Academic Publishers. doi:10.1007/978-94-017-2940-6.

Washington Post. (2012) "Rallying Troops in the Growing Cyberwar," *The Washington Post*, 13 November 2012, AA09.

Weimann, G. (2005) "How Modern Terrorism Uses the Internet," *Journal of International Security Affairs*, Spring, 8.

———. (2008) "Cyber-Terrorism: Are We Barking at the Wrong Tree?," *Harvard Asia Pacific Review*, 9, 2: 41–6.

Weiner, T. (1996) "Head of C.I.A. Plans Center to Protect Federal Computers," *The New York Times*, 26 June 1996, B7.

Weisman, S. (2015) "The Hacking of OPM: Is It Our Cyber 9/11?," *USA Today*, 13 June 2015. Online. Available: <www.usatoday.com/story/money/columnist/2015/06/13/hacking-opm-weisman/28697915/> (accessed 13 June 2015).

Weiss, J. (2011) "Water System Hack – The System Is Broken," *ControlGlobal.com*, 17 November 2011. Online. Available: <https://community.controlglobal.com/content/water-system-hack-system-broken> (accessed 17 November 2011).

White House Press Office. (2009) "Remarks by the President on Securing Our Nation's Cyber Infrastructure," *White House Press Office*, 29 May 2009. Online. Available: <www.whitehouse.gov/the_press_office/Remarks-by-the-President-on-Securing-Our-Nations-Cyber-Infrastructure/> (accessed 29 May 2009).

Whitehouse, S. (2011a) "Cybersecurity Needs Complete Plan," *Politico*, 7 March 2011a. Online. Available: <https://dyn.politico.com/printstory.cfm?uuid=8D587513-F55D-BEB0-A25C03B5B2018326> (accessed 7 March 2011a).

———. (2011b) "Whitehouse, Kyl Introduce Legislation to Enhance Cyber Security," *Whitehouse.Senate.Gov*, 11 April 2011b. Online. Available: <https://whitehouse.senate.gov/newsroom/press/release/?id=EFDEEFF7-F79B-4325-BCAD-34C8EF02A0CD> (accessed 11 April 2011b).

Whitlock, C. (2014) "Ashton Carter, Passed Over Before, Gets Picked by Obama to Be Defense Secretary," *The Washington Post*, 5 December 2014. Online. Available: <www.washingtonpost.com/world/national-security/ash-carter-passed-over-before-gets-picked-by-obama-to-lead-pentagon/2014/12/05/33a2429a-7c95-11e4-9a27-6fdbc612bff8_story.html> (accessed 5 December 2014).

Wilkins, A. (2012) "Cybergeddon," *AV Club*, 25 September 2012. Online. Available: <https://tv.avclub.com/cybergeddon-1798174299> (accessed 25 September 2012).

Witte, K. (1998) "Fear as Motivator, Fear as Inhibitor: Using the Extended Parallel Process Model to Explain Fear Appeal Success and Failure," in Anderson, P.A. and Guerrero, L.K. (eds) *Handbook of Communication and Emotion: Theory, Applications, and Contexts*, San Diego, CA: Academic Press, pp. 423–50.

Witte, K. and Allen, M. (2000) "A Meta-Analysis of Fear Appeals: Implications for Effective Public Health Campaigns," *Health Education & Behavior*, 27, 5: 591–615. doi:10.1177/109019810002700506.

Wolfe, J. (2018) "The National Intelligence Director Issued a Warning About a Cyber 9/11-Like Cyberattack," *Slate Magazine*, 19 July 2018. Online. Available: <https://slate.com/technology/2018/07/u-s-intel-chief-warns-of-a-crippling-cyberattack-against-our-critical-infrastructure-what-does-he-mean.html> (accessed 19 July 2018).

Yoran, A. (2010) "Cyberwar or Not Cyberwar? And Why That Is the Question," *Forbes.com*, 25 March 2010. Online. Available: <https://blogs.forbes.com/firewall/2010/03/25/cyberwar-or-not-cyberwar-and-why-that-is-the-question/> (accessed 25 March 2010).

Zakaria, T. (2011) "Former CIA Official Sees Terrorism-cyber Parallels," *Reuters*, 3 August 2011. Online. Available: <www.reuters.com/article/2011/08/03/us-usa-security-cyber-idUSTRE7727AJ20110803> (accessed 3 August 2011).

Zetter, K. (2009) "Lawmaker Wants 'Show of Force' Against North Korea for Website Attacks," *Wired Threat Level*, 10 July 2009. Online. Available: <www.wired.com/threatlevel/2009/07/show-of-force/> (accessed 10 July 2009).

3 From wire devils to cyber squirrels

Cyber-doom rhetoric as fear of technology-out-of-control

Introduction

In July 2016, an anonymous group took to the Internet to boast of their prowess in attacking American critical infrastructure systems with impunity. In a letter posted on the *Foreign Policy* magazine's website, the group claimed to have caused blackouts in twenty-six U.S. states in the previous month alone. They also claimed to have carried out two successful attacks on the NASDAQ, in addition to targeting schools and universities, hospitals, government buildings, airports, and military bases. They even mused about the damage that could be caused by wide-scale, coordinated cyber-physical attacks. They asked, "[W]hat if hackers had guns? Didn't you see *Skyfall*?!" Their answer: "No, we didn't. We're squirrels" (@cybersquirrel1, 2016).

In reality, an anonymous information security expert going by the handle, @cybersquirrel1, wrote the letter. This expert had gained attention for creating a map showing all of the "cyberattacks" on the power grid carried out by squirrels and other animals. The point of the project and the letter to *Foreign Policy* was to combat what he or she called the "hype" and "fear-mongering" in the U.S. public debate about cybersecurity. Cyber Squirrel noted the constant drumbeat of warnings about "cyber Armageddon" and similar doom scenarios. But, Cyber Squirrel claimed, if we factor in the hundreds of electrical blackouts caused by animals, "We experience 'armageddon' every day" (Metcalfe, 2016). So, why all the concern with hypothetical, future versions of the kind of event that occurs daily without much notice?

This chapter explores the answer to that question and, in so doing, a second reason why cyber-doom rhetoric persists. It argues that the fears expressed in cyber-doom rhetoric are not just for the immediate effects of potential cyberattacks (e.g. lack of electricity, physical damage, loss of life). Rather, fear of cyber-doom is fear of a lack of control over society, fear that forces beyond our control, in particular new technology in the hands of nefarious, intentional actors, can cripple not just infrastructure, but society itself. We saw this in Schwartau's early work, as well as in more recent examples like the *National Geographic Channel*'s 2013 "docudrama," *American Blackout*. As we recall from the previous chapter, in *American Blackout*, "keeping society running" was synonymous with functioning technological infrastructure, electricity in particular. In that fictional tale of cyber-doom, loss of electricity quickly led to social chaos and collapse, even loss of "civilization." Conversely, restoration of electricity just as quickly restored a civilized society.

As we will see in the following pages, these fears are a manifestation of deeper cultural and historical anxieties about technology and society out of alignment and out of control. In the next section, we will look at the common belief in technological determinism and the pessimism and fear it can cause. Then, we will examine the emergence of complex sociotechnical systems like electric grids and communication networks, in particular, as sources of fear during the last century. Finally, we will turn our attention to how these fears have been reflected in Western military thought. Throughout the chapter, we will draw from research in the history of technology, military history, sociology, and psychology. We will see that imagining and warning about the worst has been a common means of generating fear of, but also coping with, technological change in Western societies. Cyber-doom rhetoric is but one more example.

Technology, determinism, and pessimism

Historians of technology have written extensively about the rise of the belief in "autonomous technology" or "technological determinism" in Western societies, as well as the increasingly prominent feelings of pessimism and fear that have come along with these beliefs. As recently as the nineteenth century, the dominant view was one that saw technological innovation as the key to human progress. In Western societies, technology was commonly viewed in reverential, even religious terms, with technological progress seen as a measure of a society's development, but also its righteousness (Adas, 1989; Hughes, 2004: Ch. 2). Historian of technology Thomas Hughes demonstrates that for some nineteenth-century Americans, technology seemed to offer the ability to reclaim a lost Eden by way of a human-built second creation (Hughes, 2004: Ch. 2). Similarly, historian David Nye argues that the "technological sublime" played a central role in American history and cultural identity (Nye, 1994).

Technologies of communication and transportation, which were one and the same until the twentieth century, played a particularly important role in this story. Communication scholar and media historian James Carey goes so far as to claim that concern with communication and its associated technologies was "a central feature of American history from the outset" (Carey, 1989: 2). He notes that for American founders like John Adams and Thomas Jefferson, the Greek democratic republic was the ideal form for political life. The problem, however, was that the American founders were attempting with their Constitution to create such a republic on a scale and across geographic distances that seemed to be in direct contradiction to the requirements that Plato and Aristotle had identified for a successful republic – i.e. large enough to be self-sustaining but small enough to have a shared vision and culture and decision-making based in debate and deliberation. Adams and Jefferson both saw technologies of transportation and communication as the solution to this problem. For example, Adams had argued in *The Federalist* that transportation projects like the building of canals would facilitate the necessary "Communication between the western and Atlantic districts and between different parts of each" (Quoted in Carey, 1989: 6). As President, Jefferson acted on these ideas, announcing in 1806 a federal program to fund the building of roads and canals with this stated goal: "New channels of communication will be opened between the states, the lines of separation will disappear, their interests will be identified and their union cemented by new and indestructible ties" (Quoted in Carey, 1989: 7). As Carey puts it, for the American founders "the technology of transport and communication would make it possible to erect the vivid democracy of the Greek city-state on a continental scale." The result has been that, in American history at least, "technology is not only artifact but actor" (Carey, 1989: 7–8).

Nonetheless, these historians also note that sentiments towards technology began to shift throughout the course of the twentieth century. Many began to question both humanity's ability to control its creations, as well as the impacts of those creations. Experiences with war and new weapons were a primary driver of this shift. Michael Adas points to the damage that WWI caused to many Europeans' belief in the power of technology, rationality, and progress (Adas, 1989: 347–73). Hughes and Nye note that WWI did not have the same impact on Americans' faith in technology. Instead, Americans would have to wait for WWII, the advent of the atomic bomb and a global arms race, the failures of scientific management in Vietnam, President Lyndon Johnson's "War on Poverty," and stories about massive environmental pollution at the hands of industry for their faith in technology to be shaken (Nye, 1994; Hughes, 1998).

Changes in the characteristics of dominant technologies were also a driver of growing ambivalence and pessimism. We see a shift during the late nineteenth and twentieth centuries from individual mechanical devices created by individual inventors to large socio-technical systems created and managed by complex, geographically dispersed organizations (Marx, 1994: 244; Marx, 1997: 972–4). The locomotive and its geographically dispersed network of rails and complex supporting organizational structures is one of the earliest examples, as were the emerging electrical, telecommunications, chemical, automotive, and other industries of the late nineteenth century (Marx, 1994: 244–5). These developments fundamentally changed what we think of as "technology." Historian Leo Marx explains, "These huge systems were replacing discrete artifacts, simple tools, or devices as the characteristic material form" of technology in modern societies (Marx, 1994: 245).

Hughes has written extensively on the emergence and implications of such systems. He is therefore worth quoting at length on what constitutes such systems:

> Technological systems contain messy, complex, problem-solving components. [. . .] Among the components in technological systems are physical artifacts, such as turbogenerators, transformers, and transmission lines in electric light and power systems. Technological systems also include organizations, such as manufacturing firms, utility companies, and investment banks, and they incorporate components usually labeled scientific, such as

books, articles, and university teaching and research programs. Legislative artifacts, such as regulatory laws, can also be part of technological systems. [. . .] [N]atural resources, such as coal mines, also qualify as system artifacts.

(Hughes, 1987: 51)

In the twentieth century, large technological systems came to dominate our lives. As examples, Hughes points to the emergence of the electrical grid, the Internet, the interstate highway system, as well as large weapon systems, such as air defense, missile, and command and control systems (Hughes, 1998). In the face of such systems, Langdon Winner observers, we came to realize that "man now lives in and through technical creations" (Winner, 1977: 3) and to "entertain the vision of a postmodern society dominated by immense, overlapping, quasi-autonomous technological systems," in which society itself becomes "a meta-system of systems upon whose continuing ability to function our lives depend" (Marx, 1994: 257).

Thus, we have seen the emergence of the twin beliefs "that technology is the primary force shaping the post-modern world" (Marx, 1997: 984), but also "that somehow technology has gotten out of control and follows its own course, independent of human direction" (Winner, 1977: 13). More specifically, Hughes argues that "the degree of freedom experienced by people in the system" can be limited as the system gains "maturity and size," or what he calls "momentum." "Large systems with high momentum tend to exert a soft determinism on other systems, groups, and individuals in society" (Hughes, 1987: 54–5). In the face of this "soft determinism," Nye says that our sense of sublimity towards technology has been tempered with a tinge of ambivalence (Nye, 1994). Leo Marx goes further, arguing that an "inevitably diminished sense of human agency" in the face of large technological systems has led to an increasing sense of "technological pessimism" (Marx, 1994: 257, 238). He defines this condition as one in which we simultaneously marvel at the innovations that have made modern life possible, but also experience "a gathering sense . . . of political impotence" and "the feeling that our collective life in society is uncontrollable" as a result of our increasing dependence upon technology (Marx, 1997: 984).

But even this sentiment must be seen in a larger historical context. Deterministic ways of thinking go beyond technology. In his rhetorical history of the role that "necessity" has played in Western political discourse, Marouf Hasian notes that deterministic, or what he calls "necessitous," thought and rhetoric have played an important role in our history. This is particularly the case in foreign policy discourse and in times of war (Hasian, 2005: 8). Necessitous thinking and rhetoric, however, go slightly beyond determinism. Here "necessity" is "used to characterize both the situations *and the responses* to those situations" (Hasian, 2005: 11). Like historians of technology, Hasian notes that necessitous thinking ultimately leads to a diminished sense of human agency, causing us to overlook the role of our own decisions (or the decisions of our political leaders) in creating the challenges we face, as well as the possible avenues of response (Hasian, 2005: 24). This is an issue to which we will return in Chapter 5.

Given its prominent role in our history of thought related to politics and technology, it should come as no surprise that determinism has been a key feature of both scholarly and popular works on the "communication revolution," the "Information Age," and the emergence of a "post-industrial" or "network society." In these works, authors typically argue that technology, in this case information and communication technologies (ICTs), have exerted such a new and disruptive influence on society, economy, and culture as to change their character in a fundamental or revolutionary way. In some cases, all of human history is periodized according to the supposedly dominant technologies of a particular age. Observers vary in the degree of agency that they see humans enjoying in the face of these revolutionary technological changes. They also vary in terms of their assessment of these changes, some taking an optimistic view, others pessimistic.

Some of the most influential scholars and public intellectuals of the last sixty years produced no shortage of such work. We can think of Manuel Castells' mostly optimistic work, for example, which identifies information and knowledge working on themselves in a feedback loop as being the core of the new economy and rise of the "network society" (Castells, 2000). A decade later, Kevin Kelly, founder and former editor of the popular *Wired* magazine, took a more pessimistic tone in his work that posits the emergence of a self-reinforcing, technology-dependent society he calls the "technium" (Kelly, 2010).

But Castells and Kelly have a long list of predecessors who have engaged in the same game. None other than Norbert Wiener, the founder of cybernetics, periodized all of human history according to dominant technologies. These included the ages of the clock, the steam engine, and our current age of information, control, and automation (Wiener, 1948, 1950). Indeed, some call Wiener the founder of our Information-Age worldview (Galison, 1994: 252–3).

Others picked up on this theme, such as sociologist Daniel Bell who, in 1973, built upon this periodization to argue that humanity had passed from the agrarian age to the industrial age and was, at that time, experiencing the emergence of the post-industrial age, which would be dominated by information and communication (Bell, 1973). Similarly, James Beniger saw in technological change the roots of "the control revolution" and the "information society" (Beniger, 1986). From the 1970s through the early 1990s, Alvin and Heidi Toffler played an important role in popularizing these ideas in works such as *Future Shock*, *The Third Wave*, and *War and Anti-War* (Toffler, 1970, 1984, 1993).

Though some optimistically saw the emerging age as one of increasing control of our societies and ourselves, even perhaps to the point of transcending our biological limitations (Kurzweil, 2005), not everyone was convinced. There were always more pessimistic, but equally deterministic voices. In 1998, for example, Gene Rochlin argued that the computer revolution was, in fact, a trap. The allure of more control, he warned, would be followed by the reality of diminished human agency in relation to our own creations (Rochlin, 1998).

As James Carey argued, these competing views of the so-called information revolution are both rooted in and reflective of our own history of thinking about the role of communication in society, a fact that writers on the subject of the Internet and computing, he says, too often ignore (Carey, 2005). The fears expressed in cyber-doom rhetoric about information and communication technologies (ICTs) out of control also have their roots in this tradition of speaking about ICTs specifi-cally, but technology more generally, in deterministic, necessitous, and sometimes also pessimistic terms. That is, cyber-doom rhetoric persists, in part, because "it expresses deeper cultural anxieties" (Glassner, 1999: 208) rooted in our histories of interaction with other new technologies.

In her history of Victorian responses to the advent of electricity, for example, Linda Simon observes that "we have inherited a legacy . . . a pattern of responses to new technologies that allure [and] threaten" (Simon, 2004: 23). Similarly, Leo Marx has argued,

> Invariably people's responses to the new – to changes effected by, say, a spe-cific technical innovation – are mediated by older attitudes. Whatever their apparent spontaneity, such responses usually prove to have been shaped by significant meanings, values, and beliefs that stem from the past.
>
> (Marx, 1994: 239)

Thus, one might say that the real "soft determinism" at work here is to be found in our own predictable responses to new technologies. As post-structuralists tell us, we are constrained (but also enabled) by the networks of rhetoric, discourse, and thought in which we find ourselves enmeshed. But we also know that all such networks have their limits, that they contain gaps and antagonisms that allow for change. Thus, we will return in later chapters to the questions of why and how we might begin to change our deterministic, necessitous thinking about cybersecurity.

But before we do that, we need to examine in more detail the "pattern of responses" that has shaped our current thinking. This is because, as Marx tells us,

> It is illusory to suppose that we can isolate for analysis the immediate, direct responses to specific innovations. [. . .] [For example, a] group's responses to an instance of medical progress cannot be understood apart from the histori-cal context, or apart from the expectations generated by the belief [or disbe-lief] that modern technology is the driving force of progress.
>
> (Marx, 1994: 239)

Likewise, our fear of cyber-doom must be understood in the context of the ways that deterministic, necessitous thinking manifested itself in responses to other technologies. That will be the focus of the remaining sections of this chapter.

Fear of technology in the age of complex socio-technical systems

Many of the concerns found in contemporary cybersecurity discourse are not unique, but rather, have strong corollaries in early twentieth-century concerns about society's increasing reliance upon interdependent and seemingly fragile infrastructure systems of various types. These included the very systems implicated in fears of cyber-doom today, such as electrical grids and electronic communication networks. In both cases, we see a "pattern of response" that Marx called "technological pessimism." People marveled at the allure of these new systems and the promise they seemed to hold while simultaneously worrying about the threats that these systems might pose.

The adoption of electricity, but also the possible absence of electricity were the new power systems to fail, served as objects of fear and anxiety. Using electrification as her case study, Linda Simon has explored "the intellectual and emotional revolution that occurred as ordinary men and women, eagerly and anxiously, grappled with the greatest new idea of their time" (Simon, 2004: 23). She documents the often-contradictory response of people of the Victorian Age to the advent of electricity and its application in the areas of communication, medicine, and municipal lighting. She finds that while people adopted rather quickly the medical applications (e.g. electric shock therapy), they resisted the applications to municipal lighting and communication, such as the telegraph and, later, radio.

She resolves this seeming paradox by showing that peoples' interpretations of the meaning of electricity were rooted in existing beliefs about spiritualism and the nature of life, in particular the notion of vitalism. In her analysis, she notes that it was common for people of the late nineteenth and early twentieth centuries to see a strong link between electricity and the life-force in all creatures, as well as electricity and spirit. If electricity was central to the force of life itself, then harnessing this force for use in medicine was a positive breakthrough indeed. Thus, people were willing to submit to electricity-related treatments of the body and the spirit. In contrast, uses of electricity for communication, lighting, and powering machines were seen as vulgar and against Nature, signs of science run amok as scientists, engineers, and industrialists endowed to machines the vital life-force that flowed through humans. Indeed, misuse of electrification, it was believed, was literally a cause of illness. Simon writes,

> The very nature of the modern, electrical, industrial age was thought to cause stress leading to illness in the form of neurasthenia, a general depletion of nervous energy in the body leading to all sorts of symptoms such as fatigue, sexual dysfunction, headache, and more.
>
> (Simon, 2004: 9, 113–21)

Thus, electricity was seen at once as both the cause and cure for the ailments of modern men and women of industrial societies.

Despite these early fears of what electrification might do to the physical, mental, and spiritual health of modern men and women, the technology was eventually accepted and seen as a symbol of a society's strength and progress. Predictably, it was not long before some began to worry about what might happen to society if electrification were to fail. By the 1920s and 1930s, electrification came to be seen by many as a "symbol of vulnerability" as much as a symbol of strength (Nye, 2010: 37). This was in part tied to the development and emerging use of another new technology, the airplane. In the early decades of the twentieth century, writers of science fiction, journalists, politicians, and military professionals hypothesized that the use of aerial bombardment, in particular against the electric power grid, could allow for the crippling of technology-dependent modern societies (Konvitz, 1990; Biddle, 2002). This is a topic to which we will return in the next section.

Early forms of electronic communication, including the radio, telegraph, and telephone, also sparked fear and anxiety by government officials and the public alike that are similar to contemporary concerns about cybersecurity. Along with transcontinental railroad networks, these "new networks of long-distance communication," which could not be "wholly experienced or truly seen," were the first of the kind of large, complex, nation-spanning, socio-technical systems that were at the heart of the last century's increasing technological pessimism (MacDougall, 2006: 720). The new communication networks were often portrayed in popular media as constituting a new space, a separate world dominated by crime, daring, and intrigue (MacDougall, 2006: 720–1). While the new communication network "gave new powers to its users, [it] also compounded the ability of distant people and events to affect those users' lives" (MacDougall, 2006: 718). In short, it introduced the power and danger of "action at a distance – the ability to act in one place and affect the lives of people in another" (MacDougall, 2006: 721). Many worried that the combination of action at a distance and the relative anonymity offered by the new communication networks would allow people to more readily engage in immoral activities like gambling, that the networks would become tools of organized crime, and even that nefarious "wire devils" – who today we would call hackers – could use the telegraph to crash the entire U.S. economy (MacDougall, 2006: 724–6). Even if particular nefarious actors could not be identified, the mere fact of a "complex interdependence of technology, agriculture, and national finance" that was difficult if not impossible to apprehend was itself enough to cause anxiety (MacDougall, 2006: 724).

There was also growing concern that the new communication technologies might not deliver on Madison and Jefferson's vision for their role in American democracy, that they might, in fact, undermine rather than foster the sense of shared identity, culture, and purpose that the founders had said was essential to the success of the American experiment. In the late nineteenth century, for example, governments (including the U.S. government) began to worry about the emergence of national news organizations capable of disseminating their content to foreign audiences via print and telegraph. This, officials feared, might undermine national unity. These concerns contributed to the creation of a period of intense international competition over who would control the physical infrastructure of this "Victorian Internet" (Standage, 1998). These concerns only intensified through the early twentieth century as the technologies of mass communication came to include not just print and telegraph, but radio and telephone too. The First World War heightened concern over the possibility that foreign enemies might exploit the new technologies of mass communication to undermine the national will during wartime. In turn, these concerns led to some of the earliest academic research on the promise and peril of propaganda techniques combined with the new technologies of mass communication and emerging knowledge from psychological sciences. Though interest waned during the interwar years, this research would resume during WWII and would spark the creation of the new field of communication studies in the early years of the Cold War (Simpson, 1994; Mattelart, 2000; Butsch, 2008; Goodman, 2011; Auerbach, 2015).

This pattern of fear and anxiety directed at new technologies continued through the twentieth and into the twenty-first century and became a subject of interest for social scientists. Like historian Leo Marx, many sociologists also see technology as central to our shift to a postmodern society (Lyotard, 1984; Castells, 2000), which can include a pessimistic or at least ambivalent view of technology's impacts. We see this, for example, in the "risk society" thesis advanced by Anthony Giddens and Ulrich Beck, where the dangers increasingly facing societies are "manufactured" as the product of our own technologies (Giddens, 1990; Beck, 1992, 1999). Similarly, Furedi observes that much of what we fear in our culture and politics of fear is related to technology (Furedi, 2006: 41–2). For others, technology, ICTs in particular, has increasingly become a focus of research on the phenomenon of "moral panic" (Krinsky, 2013: 227–84; Wall, 2008), with one researcher coining the term "technopanic" to describe this subgenre of the moral panic phenomenon (Marwick, 2008; for an application of the concept to cybersecurity, see Thierer, 2013). Finally, psychologists have coined the term "technophobia" to describe the sometimes abnormal or irrational fear of technology, ICTs in particular, that can sometimes plague individuals, but also groups or entire societies. Technophobia, they observe, can be evident in individual thought and behavior, but also expressed culturally through works of popular fiction like film and literature, especially science fiction (Brosnan, 1998; Dinello, 2005).

Socio-technical change as antagonism in military thought

Military thinkers have not been immune to these tendencies towards deterministic thinking about, and ambivalence towards, technology. We tend now to think of the military as leading the way in creating and adopting new technologies, especially in the area of ICTs. After all, we are reminded that the military first created the Internet. But militaries have not always had such a relationship with technology. What's more, there has been, and still is, a greater degree of military ambivalence towards ICTs than most have recognized (Lawson, 2013). The fears and anxieties expressed in cyber-doom rhetoric are shaped in part by, but also help to shape, this larger institutional relationship to new technologies.

In this section, we will examine how wider cultural fears and anxieties about large technological systems of the late nineteenth and early twentieth centuries were reflected in military thought of that time. We then turn our attention to how these patterns have played out with respect to military adoption of ICTs from World War II up to the "war on terrorism." Finally, we will take special note of the roles that projection, as well as the use of fiction, have had in military thinking about new technologies and the future of warfare.

In theories of industrial-mechanized warfare from the early twentieth century, as well as later theories of Information-Age warfare, we see a tendency towards determinism and ambivalence. In both cases, military theorists have seen technological change as the driver of massive changes in the wider socioeconomic system. They have worried that military thought, practice, and technology are out of sync with these changes. They have seen technology as simultaneously the new driving force allowing for new capabilities, but also as a force creating frightening new vulnerabilities that could spell doom (Lawson, 2011a: 44).

In the early twentieth century, mechanization was driven as much by fear of new technology as it was by excitement over its promise. In turn, these fears were reflective of a more generalized anxiety about the supposed interdependence and fragility of mòdern, industrial societies. This anxiety shaped the thinking of military planners on both sides of the Atlantic. Like cyber war theorists today, mechanized warfare theorists argued that the unique vulnerabilities resulting from society's newfound dependence on technology – at that time, interlocking webs of production, transportation, and communication systems – could be exploited to cause almost instantaneous chaos, panic, and paralysis in a society (Biddle, 2002; Konvitz, 1990).

Early airpower theorists had these assumptions at the heart of their plans for the use of strategic bombardment. For example, in his influential 1925 book, *Paris, or the Future of War*, B.H. Liddell Hart argued, "A modern state is such a complex and interdependent fabric that it offers a target highly sensitive to a sudden and overwhelming blow from the air" (Liddell Hart, 1925: 41). He continued, "A nation's nerve-system, no longer covered by the flesh of its troops, is now laid bare to attack, and, like the human nerves, the progress of civilization has rendered it far more sensitive than in earlier and more primitive times" (Liddell Hart, 1925: 37). Thus, he argued simultaneously that "in the mechanical future of war supremacy will go to the nation with the greatest industrial resources," but also that interdependence and fragility meant that "in a modern nation at war its industrial resources and communications form its Achilles' heel" (Liddell Hart, 1925: 52–3, 33–4).

Similarly, in the United States, Major William C. Sherman, who co-authored the 1922 Air Tactics text used to train American pilots, believed industrialization to be both a blessing and a curse. On the one hand, industry was at the very heart of the modern nation's ability to wage war successfully. But on the other, he believed that a massive strike against an enemy's "key plants" could have the effect of delivering a knock-out blow. His "industrial fabric" theory of aerial bombardment started from the assumption that the "very quality of modern industry renders it vulnerable" to aerial attack (Sherman, 1926: 217–18).

As we saw in the previous section, in the early twentieth century, many believed that industrialization itself was making society ill through the depletion of nervous energy and creation of nervous disorders. These fears also made their way into theories of mechanized warfare. Derived from popular writings about crowd psychology, eugenics, and social Darwinism, military theorists identified a second vulnerability of industrialized societies: industrialization-induced changes in human physiology resulting in the "nervous complexion of the modern mind" (Biddle, 2002: 14, 18). The urban poor and working classes, often referred to as "industrial populations," were seen as particularly prone to the maladies of neuroses, panic, and hysteria (Biddle, 2002: 13, 64; Clodfelter, 2010; Freedman, 2005; Jones et al., 2006). Liddell Hart argued that bombardment of "the slum districts" would result in "industrial populations" being "maddened into the impulse to maraud" (Liddell Hart, 1925: 41–2). Sherman believed that the "elbow-to-elbow" living conditions of industrial populations made them prone to the transmission of a "wave of hysteria" and the "propagation of panic," both of which he considered a "disease" (Sherman, 1926: 13–14).

But these fears did not drive military planners away from adoption of the new technologies and the weapons that they enabled. Instead, the "prophets of mechanization," as historian Bart Hacker has called them, pushed ahead with the adoption of the technologies behind industrialization, including weapons like tanks and airplanes powered by internal combustion engines and mass-produced on factory assembly lines (Hacker, 1982: 56). Their plan for using those weapons would seek to take advantage of the supposed lessons about the "nervous complexion" of "industrial populations." That plan was the targeting of civilian populations first and foremost. Liddell Hart, Sherman, and many proponents of strategic bombing viewed it as "an aerial shortcut" to defeating the enemy's will to fight (Biddle, 2002: 139–40). Given their supposed susceptibility to panic, neuroses, and hysteria, the bombardment of civilian populations was seen as the fastest, most effective means to that end (Liddell Hart, 1925: 31, 37; Sherman, 1926: 5, 210). When these theories were put into action during World War II, Biddle notes that the result was "nothing less than a form of aerial Armageddon played out over the skies of Germany and Japan" (Biddle, 2002: 9). We will return in the next chapter to the question of whether these tactics actually caused the panic and paralysis in the enemy that they were intended to cause.

Military fear of new technology did not end with mechanization, however, but extended to the new technologies of electronic communication and reflected the broader concerns mentioned in the last section. Some worried that the new communication technologies could be used not just against society generally, but also against the military directly. The U.S. Navy, for example, was initially reluctant to adopt the radio, in part because of concern over what today would be called "information assurance" (Douglas, 1985). The early twentieth century saw an explosion in the number of amateur radio users in the United States who could not only "listen in" on military radio traffic, but who could also broadcast on the same frequencies used by the military. Amateur broadcasts could clog the airwaves, preventing legitimate military communications, but could also be used to feed false information to ships at sea. In response, the Navy worked to have amateurs banned from the airwaves. They succeeded only in 1912 after it was reported that interference by amateur radio operators may have hampered efforts to rescue survivors of the Titanic disaster. After 1912, amateurs were limited to the shortwave area of the electromagnetic spectrum and during World War I, the U.S. government banned amateur radio broadcast entirely (Douglas, 2007: 214–15).

As mentioned earlier, many historians trace the beginning of the Information Age to World War II and the immediate postwar period. We can also trace the beginnings of Information-Age weapons and forms of warfare to this period. Of course, advanced science and technology were central to the Allies' victory in that war and, largely as a result, have been increasingly important to warfare ever since, especially in the United States. Nonetheless, the U.S. military's adoption of weapons and strategies of Information-Age warfare have been propelled as much by fear of being out of sync with seemingly autonomous and revolutionary technological forces as it has by dreams of future warfare finally free of Clausewitzian fog and fiction. Along the way, there have always be skeptics of the new technologies, even among some of those most responsible for the doctrines and strategies that have shaped the U.S. way or warfare over the last sixty years. In the next several pages, we will examine the roles of fear and skepticism of technology in the emergence of Information-Age warfare.

The general purpose digital computer and methods of analysis reliant upon it, the so-called "systems sciences," emerged in the U.S and the U.K in the years prior to the outbreak of WWII. These tools played an important role during the war, from planning bombing missions, to air defense, to anti-submarine warfare, and more. In the years following the war, each of the military services in the United States set up "study and analysis" centers like the Air Force's RAND Corporation to further refine the tools of systems science and computing as applied to warfare. Even from the start, however, there was an element of ambivalence at work. On the one hand, fear of the implications of revolutionary new technology-out-of-control served as an impetus for the creation and adoption of systems science and computing. The demonstrated effects of atomic weapons at the close of the war, combined with no other history of these new weapons' use to guide planning and decision-making, meant that military professionals and civilian defense intellectuals would increasingly rely on another new technology, the digital computer, for understanding, commanding, and controlling war in the atomic age (Fortun and Schweber, 1993; Ghamari-Tabrizi, 2000, 2005; Lawson, 2011b).

Simultaneously, however, the introduction and spread of systems science and computers in the military led to serious criticism. Critics charged that the methods and tools of systems science were inappropriate for producing militarily relevant knowledge, that they were a threat to the professional identities and authority of those in uniform, and that they led to increased organizational centralization, which they said resulted in a less effective military. Indeed, some of the most influential military voices of the era were harsh critics of attempts to turn warfare into a scientific endeavor through the use of computers and systems science. This included none other than Gen. Curtis LeMay, the USAF General who had founded the Strategic Air Command and later served as Air Force Chief of Staff. Ironically, he had also played a role in the founding of RAND (Lawson, 2011b; LeMay and Smith, 1968).

The Vietnam experience provided ammunition for the critics of military computerization and use of systems science, both of which critics saw at the heart of U.S. failures in that war. Nonetheless, the acceleration of computerization was at the heart of post-Vietnam military reform efforts in the United States. The so-called Offset Strategy was meant to address the twin problems of Soviet nuclear parity and quantitative superiority in conventional weapons. This would be achieved in two ways. First was the adoption of ICT-enabled weapon systems, such as intelligence; surveillance and reconnaissance systems; precision-guided weapons; and digital command, control, and communications systems (Tomes, 2007).

Second was the adoption of a new set of service-level doctrines for countering the perceived Soviet threat. The doctrine that ultimately emerged was called "maneuver warfare." The Army and Marine Corps adopted this doctrine in 1982 and 1989, respectively (Department of the Army, 1982; Romjue, 1984; United States Marine Corps, 1989). Surprisingly, however, some of the key authors and promoters of this doctrine were harsh critics of the new technologies being developed as part of the Offset Strategy. This included, most notably, retired Air Force Colonel John Boyd and his acolytes. This self-styled "military reform movement" played a central role in the development of both the Army and Marine Corps maneuver warfare doctrines. But they were also harsh critics of the use of systems science, computerized war gaming, command and control systems, and even communications systems (Lawson, 2014: 81–3). For example, in their 1986 book, *America Can Win*, Senator Gary Hart (D-CO) and his aide, William Lind, advocated that the military should rely on "personal, face-to-face meetings" and "not teletyped messages printed on computer paper"; on "implicit communications" and "a shared way of thinking" instead of "fusion centers and cathode-ray tubes." Of most communication technologies, they said, "The best thing a combat unit could do with most of it is to kick it off the back of the truck when they move out to fight" (Hart and Lind, 1986: 52–3).

These were not insignificant voices. Hart was a U.S. Senator and one-time presidential candidate who had co-founded the Military Reform Caucus in Congress in 1981 (Mohr, 1984). Lind had helped to kick off a massive debate about U.S. military reform efforts with a series of articles in 1979 (Lind, 1979a, b). What's more, Lind had lectured at the Marine Corp's Amphibious Warfare School, where he taught Boyd's theories of warfare to Marine officers. He even wrote his own book on maneuver warfare (Lind, 1985). He and Boyd both later contributed directly to the Marines' adoption of maneuver warfare (For more on Lind, Boyd, and 1980s military reform, see Chapter 4 in Lawson, 2014).

The world witnessed the implications of these technological and doctrinal reforms in January 1991 during Operation Desert Storm in Iraq. Indeed, U.S. military professionals seemed just as amazed as many outside observers at the rapid U.S. victory over Iraqi forces. Even as the war was ongoing, the Office of Net Assessment (ONA) in the Pentagon began undertaking a study of the "military-technical revolution" (later re-named "revolution in military affairs" [RMA]) brought about by the combination of ICTs, precision weapons, and new doctrines. Ironically, though the U.S. purposefully launched this "revolution" in the late 1970s and 1980s with its Offset Strategy, its Soviet rival was first to notice and theorize the revolutionary potential of these changes. The ONA report was meant to assess the degree to which the Soviet observations of the U.S. reforms were correct, to assess whether they really were a revolution. The report concluded that they were (Krepinevich, 1992).

Nonetheless, technological determinism and pessimism would both feature prominently in the debates about Information-Age warfare that ensued in the following decade and a half. By 1994, one of the authors of the ONA report raised concerns about "keeping pace with the military-technical revolution," as though it were an independent force and not, in fact, a revolution of our own making (Krepinevich, 1994). Without an obvious "peer competitor" at the end of the Cold War, many came to see the Information Age itself as the primary challenge. This tendency was reinforced by the works of futurists Alvin and Heidi Toffler, whose popularizations of the deterministic strain of Information-Age theorizing was mentioned in earlier chapters. Like Wiener and others before them, the Tofflers had divided all of human history into three periods (or "waves"). Each wave was defined by the dominant technology of that time period. In their technological determinist view, economies, societies, and "war-forms" were all determined by the dominant technology (For more on the Tofflers' thinking and impact in the U.S. military, see Lawson, 2014: 103–4). In the mid-1990s, Chief of Staff of the Army General Gordon R. Sullivan drew extensively from the Tofflers' work, including when expressing the ambivalent position that with the demise of the Soviet Union, the Information Age itself "present[s] us with both an unprecedented challenge and an unparalleled opportunity" (Sullivan and Coroalles, 1995: 3). In the U.S. Army, at least, the new conventional wisdom of the 1990s was that the world of the Information Age was newly volatile, uncertain, complex, ambiguous, and dangerous (Stiehm, 1996; Sullivan and Dubik, 1993).

As I have documented in detail elsewhere (Lawson, 2014), ideas about technology-out-of-control or running ahead of our ability to keep pace and leading to a more dangerous world were at the heart of the theories and strategies of warfare that took the United States to war in Iraq in March 2003. Importantly, they were also foundational assumptions of the strategies and doctrines of counterinsurgency implemented in 2007 in an attempt to snatch victory from the jaws of defeat when U.S. civilian and military leaders realized that the initial plan had failed to deliver victory.

The notion of "network-centric warfare" (NCW), most notably as articulated by VADM Arthur Cebrowski and various of his co-authors from 1997 to 2003, played an important role in shaping U.S. thinking in the run-up to the 2003 U.S. invasion of Iraq. As one NCW theorist argued in the week prior to the U.S. invasion, these theories explained "why we're going to war, and why we'll keep going to war" (Barnett, 2003). Like earlier notions of RMA from the 1990s, NCW theorists were influenced to a great degree by the Tofflers' determinist theories of sociotechnical change. They argued that economies and societies had been fundamentally changed as a result of the revolution in ICTs, with several dangerous implications for the military.

The first implication was that some economies had not yet integrated into the new, globalized economy. These "non-integrating gaps" of globalization, NCW theorists believed, were the source of dangerous new threats like the terrorist attacks of

September 11, 2001. Second, they argued that the U.S. military had not, in fact, kept pace with a sociotechnical revolution that had enabled adversaries like al-Qa'ida to carry out their attacks. Thus, NCW theorists argued that the U.S. military must redouble its efforts to adopt the same technologies (i.e. ICTs) and organizational forms (i.e. networked, decentralized, dispersed) as its adversaries. What's more, these ideas of sociotechnical-change-out-of-control were used to justify the use of preventive force to carry out regime change in the "non-integrating gaps" of globalization and against "super-empowered individuals and groups" worldwide (i.e. targeted killing, special forces raids, drone strikes, etc.) (Lawson, 2014: 118–25).

By 2004, military professionals themselves, especially in the Army and Marine Corps, were beginning to recognize that the U.S. was not succeeding in countering a growing insurgency in Iraq. They began to develop an alternative, counterinsurgency (COIN) approach to the fight. This approach was enshrined as official doctrine in 2006 and adopted as the official U.S. strategy in Iraq in 2007 (Department of the Army, 2006; Ricks, 2009b).

Even though COIN was meant to reverse the failures of the previous approach, it was still rooted in the same determinist assumptions about the supposed relationship between technological, social, and economic change found in NCW and RMA. Like these earlier theorists, the so-called COINdinistas (Ricks, 2009a) also saw technological change and economic globalization at the heart of the new threats facing the United States. One key COINdinista, Col. David Kilcullen, who helped author the joint Army/Marine Corps doctrine manual on COIN, wrote that al-Qa'ida had been uniquely empowered by "globalized communications, finances and technology" (Kilcullen, 2004: 16). These ideas were reflected in the official COIN doctrine, which said that America's enemies employed a combination of "modern technology with ancient techniques of insurgency and terrorism" (Department of the Army, 2006: ix). The manual explained: "Interconnectedness and information technology are new aspects of this contemporary wave of insurgencies. Using the Internet, insurgents can now link virtually with allied groups throughout a state, a region, and even the entire world" (Department of the Army, 2006: 1–4).

As with RMA and NCW, however, the new ICTs were seen as both a curse and a blessing. In fact, in the post-9/11 era of U.S. global counterinsurgency, the U.S. military has had an ambivalent view of ICTs and their role in conflict. Of course, as we have just seen, ICTs are viewed as a unique enabler of America's enemies, first al-Qa'ida and more recently the Islamic State (ISIS). On the other hand, like NCW theorists before them, COINdinistas have nonetheless argued that the U.S. military has no choice but to get in sync with these larger changes by adopting the technologies, strategies, and organizational structures of the Information Age if it is to succeed in its fight against these groups (Lawson, 2014: 144–5). And yet, even as U.S. leadership in Iraq moved to expand the U.S. military's use of ICTs, in particular social media, in its battle for both Iraqi and American "hearts and minds," others within the U.S. defense community warned of the possible dangers that social media might pose, including to military operational security (OPSEC) and cybersecurity (Lawson, 2013).

In the discourses of NCW and COIN, we also begin to see a hint of projection as we defined it in the last chapter, which was seeing in current or potential adversaries the very actions we have already taken or wish to take. Indeed, projection played an important role in debates about the futures of industrial- and Information-Age warfare, respectively. It has taken many of the same, particular forms as well. These can include warnings that current or potential adversaries have actually overtaken us in adoption of new technologies, thus posing a threat and necessitating that we play catch-up. Of course, in many cases, these warnings are exaggerations at best, or outright fabrications at worst.

One can point to numerous examples. One might think of the dire warnings about supposed bomber or missile gaps during the Cold War, from the early days of the 1960s all the way into the 1980s on the eve of the Soviet Union's collapse. In both cases, as we now know, these "gaps" were exaggerated or non-existent (Oreskes and Conway, 2010). More recently, we can point to erroneous claims of an Iraqi WMD threat to the United States as the primary justification for the 2003 invasion of Iraq (Mitchell, 2006; Thrall and Cramer, 2009). Perhaps it might have made sense to worry that other nations, particularly a super power rival, would overtake the United States in certain areas of technology in which the latter had enjoyed an advantage. But non-state actors like terrorist groups have also been the object of U.S. projection.

As mentioned earlier, NCW theorists argued that the United States had not, in fact, kept pace with the military-technical revolution. Globally networked ICTs, they said, had enabled the emergence of transnational, non-state actors such as al-Qa'ida, who Cebrowski and Barnett called "super-empowered individuals" because of their ability to wield power and to have effects at a level once reserved for states (Cebrowski and Barnett, 2003). Secretary of Defense Donald Rumsfeld went so far as to warn,

> In an age when terrorists move information at the speed of an email, money at the speed of a wire transfer, and people at the speed of a commercial jetliner, the Defense Department is bogged down in the micromanagement and bureaucratic processes of the industrial age – not the information age.
>
> (Quoted in Cebrowski, 2003)

In short, not only had the United States not kept pace with the military-technical revolution, even groups like al-Qa'ida were perceived to have moved ahead of the U.S. military. Thus, Cenbrowski and Barnett said that the U.S. military must "morph . . . to mirror the target set" (Cebrowski, 2003), that is, become more like al-Qa'ida, become "a military of super-empowered individuals [capable of] fighting wars against super-empowered individuals" (Cebrowski and Barnett, 2003).

Once again, influential COINdinistas shared these same assumptions. For example, David Kilcullen worried that the U.S. military was stuck in the industrial age while its adversaries had made the transition to the Information Age. They would have found little disagreement among the NCW theorists in saying that "the industrial-age approach is inadequate for today's conflict environment"

(Kilcullen, 2005: 415) and that "existing organizations and concepts are ill-suited" to this new era of conflict (Kilcullen, 2003: 41).

Finally, use of fiction as a tool for thinking about new technologies and what they might mean for the future of warfare is not unique to cyber conflict. There has been a symbiotic relationship over the years between various forms of fiction, including games, simulations, scenarios, and literature, and the development of new weapons technologies and the strategies for fighting future wars with them. As we saw in Chapter 2 with cyber warfare, sometimes defense professionals create these fictions, while in others cases they are influenced by fictions imported from the civilian world. We can see examples of this reciprocal relationship in the cases of literature, as well as gaming and simulation.

In the case of literature, the British English professor I.F. Clarke described in detail the emergence of what we would now call military science fiction beginning in the eighteenth century. These tales of the next war were often inspired by real life scientific, technological, or military events – e.g. new scientific discoveries, development of new technologies, or stunning battlefield victories or defeats. This genre of literature became especially popular in Europe and the United States beginning in the late nineteenth century and through the early years of the twentieth century preceding the onset of World War I (Clarke, 1966, 1995). But it was not just authors of novels and short stories who were influenced by events in science, technology, and military affairs. These tales of future wars waged with frightening new weapons shaped the public imagination in turn, including the imaginations of politicians and military leaders, turning those stories into "a material force" (Franklin, 2001: 5). We see an example of the material force of such tales in the writings of H.G. Wells, whose stories about the future of air warfare in particular influenced military theorists on both side of the Atlantic in the years leading up to World War I, including those of B.H. Liddell Hart discussed earlier (Travers, 1975; Biddle, 2002).

We find a similar historical relationship between gaming and simulation on one hand and the military and warfare on the other. Of course, some of our oldest games can trace their origins to warfare. This is the case, for example, with board games like chess and go (Halter, 2006). The more recent advent of computer simulation and gaming is also tied closely to military history. As mentioned earlier, the general purpose digital computer was developed in the context of World War II and further refined during the Cold War where it became central to carrying out the "systems sciences" increasingly applied to the study of future warfare by organizations like RAND. Table-top role playing games and computer simulations emerged as important methods for trying to understand what might happen in a future war conducted with atomic weapons, across global distances, and at supersonic speeds (Ghamari-Tabrizi, 2000). Computerized systems analysis and simulation made an important contribution to the development and adoption of the maneuver warfare theory and doctrine mentioned earlier. Studies using these tools helped to demonstrate the need for the new doctrine (Romjue, 1984; Starry, 1978), but they were also used to train military professionals on the use of a whole new generation of advanced weapons systems meant to implement the doctrine and that were being procured as part of the so-called Reagan build-up. This, in turn, set off a reciprocal relationship of influence between the military and the entertainment industry, including the burgeoning videogame industry, that many now refer to as the "military-entertainment complex" (Lenoir, 2000, 2002, 2003; Ghamari-Tabrizi, 2004; Der Derian, 2009).

Conclusion

In the preceding pages, we explored a second reason why cyber-doom rhetoric persists in U.S. public policy discourse. That is, not only is it exemplary of a wider culture and politics of fear in contemporary American society, but it is also reflective of deep-seated fears and anxieties about technology more generally in America history. As we have seen, America's founders viewed communication and transportation technologies as infrastructures critical to binding the new nation together, for developing a shared sense of vision, culture, and purpose, and for carrying out the deliberative and decision-making functions necessary to governing a people spread over vast geographic distances. We might say, then, that the idea that functioning infrastructure and society are one and the same, as often expressed in cyber-doom scenarios, has been with us from the beginning. We have not just had a tendency to see technology as a primary driver of history, but as the glue that holds our society together. This technological determinist view can lead to fear and anxiety that were those technologies to fail, so would our society. Our fear of technology-out-of-control is, at bottom, a fear of society-out-of-control. This anxiety has been exacerbated by a shift in the dominant characteristics of technology from individual tools to globe-spanning networks that are difficult to apprehend and which offer the promise and peril of action at a distance. Networks of electronic communication, from the telegraph to the Internet, are exemplary of such large technological systems that have been of increasing concern over the last century. As discussed earlier, these concerns have been at the heart of thinking about the future of warfare and national security during this same period. Much like our fears of cyber-doom today, military theorists of the last 150 years worried about increasing societal and economic dependence on vast technological infrastructures and the supposedly inevitable paralysis that would result if they were to fail. They worried about adversaries developing and using against us the capabilities we had already developed or used against others. They turned to fictions of various kinds, from games and simulations to literature and film, to help them think about what revolutionary new technologies might mean for the future of warfare.

But in doing so, Lawrence Freedman has demonstrated that they were often wrong in their predictions and sometimes with disastrous results (Freedman, 2005, 2017). In the next chapter, we will turn our attention to the question of what has really happened when infrastructures have failed either from human or natural causes. Prophets of cyber-doom, like the prophets of mechanization before them, predict the inevitable collapse of society as a result of infrastructure failure. As we saw in Chapter 2, they point to historical and contemporary incidents of war, terrorism, or natural disaster, as well as cyberattacks that "could have been worse," to bolster the case for cyber-doom. But as we will see, even the worst of the events to which the prophets of cyber-doom point have not caused the kind of panic, paralysis, and social and economic collapse contemplated in cyber-doom scenarios.

References

Adas, M. (1989) *Machines as the Measure of Men: Science, Technology, and Ideologies of Western Dominance*, Ithaca: Cornell University Press. doi:10.7591/9780801455261.

Auerbach, J. (2015) *Weapons of Democracy: Propaganda, Progressivism, and American Public Opinion*, Baltimore: Johns Hopkins University Press.

Barnett, T.P.M. (2003) "The Pentagon's New Map," *Esquire*, 1 March 2003. Online. Available: <www.esquire.com/ESQ0303-MAR_WARPRIMER?click=main_sr> (accessed 1 March 2003).

Beck, U. (1992) *Risk Society: Towards a New Modernity*, London: Sage Publications.

———. (1999) *World Risk Society*, Malden, MA: Polity Press.

Bell, D. (1973) *The Coming of Post-Industrial Society: A Venture in Social Forecasting*, New York: Basic Books.

Beniger, J.R. (1986) *The Control Revolution: Technological and Economic Origins of the Information Society*, Cambridge, MA: Harvard University Press.

Biddle, T.D. (2002) *Rhetoric and Reality in Air Warfare: The Evolution of British and American Ideas About Strategic Bombing, 1914–1945*, Princeton, NJ: Princeton University Press. doi:10.1515/9781400824977.

Brosnan, M.J. (1998) *Technophobia: The Psychological Impact of Information Technology*, New York: Routledge. doi:10.4324/9780203436707.

Butsch, R. (2008) *The Citizen Audience: Crowds, Publics, and Individuals*, New York: Routledge. doi:10.4324/9780203929032.

Carey, J.W. (1989) *Communication as Culture: Essays on Media and Society*, New York: Routledge. doi:10.4324/9780203928912.

———. (2005) "Historical Pragmatism and the Internet," *New Media & Society*, 7, 4: 443–55. doi:10.1177/1461444805054107.

Castells, M. (2000) *The Rise of the Network Society*, Oxford: Blackwell Publishers. doi:10.1002/9781444319514.

Cebrowski, A.K. (2003) *Network Centric Warfare and Transformation*, Presented at IDGA Network Centric Warfare Conference, Arlington, VA, 22 January 2003.

Cebrowski, A.K. and Barnett, T.P.M. (2003) "The American Way of War," *Proceedings of the U.S. Naval Institute*, 129, 1: 42–3.

Clarke, I.F. (1966) *Voices Prophesying War, 1763–1984*, London: Oxford University Press.

———. (1995) *The Tale of the Next Great War, 1871–1914: Fictions of Future Warfare and of Battles Still-to-come*, Syracuse, NY: Syracuse University Press.

Clodfelter, M. (2010) "Aiming to Break Will: America's World War II Bombing of German Morale and Its Ramifications," *Journal of Strategic Studies*, 33, 3: 401–35. doi:10.1080/01402390903189436.

Department of the Army. (1982) *Field Manual 100–5, Operations*, Washington, DC: Department of the Army.

Department of the Army. (2006) *FM 3–24: Counterinsurgency*, Washington, DC: Headquarters, Department of the Army.

Der Derian, J. (2009) *Virtuous War: Mapping the Military-Industrial-media-entertainment Network*, New York: Routledge. doi:10.4324/9780203881538.

Dinello, D. (2005) *Technophobia!: Science Fiction Visions of Posthuman Technology*, Austin: University of Texas Press.

Douglas, S.J. (1985) "Technological Innovation and Organizational Change: The Navy's Adoption of Radio, 1899–1919," in *Military Enterprise and Technological Change: Perspectives on the American Experience*, Cambridge, MA: MIT Press, pp. 117–74.

———. (2007) "Early Radio," in Crowley, D. and Heyer, P. (eds) *Communication in History: Technology, Culture, Society*, Boston: Pearson, pp. 210–16. doi:10.4324/9781315664538.

Fortun, M. and Schweber, S.S. (1993) "Scientists and the Legacy of World War II: The Case of Operations Research," *Social Studies of Science*, 23, 4: 595–642. doi:10.1177/030631293023004001.

Franklin, B. (2001) *War Stars: The Superweapon and the American Imagination*, London: Oxford University Press.

Freedman, L. (2005) "Strategic Terror and Amateur Psychology," *The Political Quarterly*, 76, 2: 161–70. doi:10.1111/j.1467-923x.2005.00668.x.

———. (2017) *The Future of War: A History*, New York: PublicAffairs.

Furedi, F. (2006) *Culture of Fear Revisited: Risk-Taking and the Morality of Low Expectation*, London: Continuum.

Galison, P.L. (1994) "The Ontology of the Enemy: Norbert Wiener and the Cybernetic Vision," *Critical Inquiry*, Autumn: 228–66. doi:10.1086/448747.

Ghamari-Tabrizi, S. (2000) "Simulating the Unthinkable: Gaming Future War in the 1950s and 1960s," *Social Studies of Science*, 30, 2: 163–223. doi:10.1177/030631200030002001.

———. (2004) "The Convergence of the Pentagon and Hollywood: The Next Generation of Military Training Simulations," in Rabinowitz, L. and Geil, A. (eds) *Memory Bytes: History, Technology, and Digital Culture*, Durham, NC: Duke University Press, pp. 150–73. doi:10.1215/9780822385691.

———. (2005) *The Worlds of Herman Kahn: The Intuitive Science of Thermonuclear War*, Cambridge, MA: Harvard University Press. doi:10.4159/9780674037564.

Giddens, A. (1990) *The Consequences of Modernity*, Stanford, CA: Stanford University Press.

Glassner, B. (1999) *The Culture of Fear: Why Americans Are Afraid of the Wrong Things*, New York, NY: Basic Books.

Goodman, D. (2011) *Radio's Civic Ambition: American Broadcasting and Democracy in the 1930s*, New York, NY: Oxford University Press. doi:10.1093/acprof:oso/9780195394085.001.0001.

Hacker, B.C. (1982) "Imaginations in Thrall: The Social Psychology of Military Mechanization, 1919–1939," *Parameters*, 12: 50–61.

Halter, E. (2006) *From Sun Tzu to Xbox: War and Video Games*, New York, NY: Thunder's Mouth Press.

Hart, G. and Lind, W.S. (1986) *America Can Win: The Case for Military Reform*, Bethesda, MD: Adler & Adler.

Hasian, M., Jr. (2005) *In the Name of Necessity: Military Tribunals and the Loss of American Civil Liberties*, Tuscaloosa: University of Alabama Press.

Hughes, T.P. (1987) "The Evolution of Large Technological Systems," in Bijker, W.E., Hughes, T.P. and Pinch, T.J. (eds) *The Social Construction of Technological Systems: New Directions in the Sociology and History of Technology*, Cambridge, MA: MIT Press, pp. 51–82.

————. (1998) *Rescuing Prometheus: Four Monumental Projects That Changed the Modern World*, New York: Pantheon Books.

————. (2004) *Human-Built World : How to Think About Technology and Culture*, Chicago: University of Chicago Press.

Jones, E. et al. (2006) "Public Panic and Morale: Second World War Civilian Responses Re-Examined in the Light of the Current Anti-Terrorist Campaign," *Journal of Risk Research*, 9, 1: 57–73. doi:10.1080/13669870500289005.

Kelly, K. (2010) *What Technology Wants*, New York: Viking.

Kilcullen, D. (2003) *Complex Warfighting*, Canberra, Australia: Australian Army Future Land Operational Concept.

————. (2004) "Countering Global Insurgency (Version 2.2)," unpublished manuscript.

————. (2005) "Complex Irregular Warfare: The Face of Contemporary Conflict," *The Military Balance*, 105, 1: 411–20. doi:10.1080/04597220500387712.

Konvitz, J.W. (1990) "Why Cities Don't Die: The Surprising Lessons of Precision Bombing in World War Ii and Vietnam," *American Heritage Invention & Technology Magazine*, 5, 3: 58–63.

Krepinevich, J. and Andrew, F. (1992) *The Military-Technical Revolution: A Preliminary Assessment*, Washington, DC: Office of Net Assessment, Department of Defense.

————. (1994) "Keeping Pace With the Military-Technological Revolution," *Issues in Science and Technology*, 10, Summer: 23–9.

Krinsky, C. (ed) (2013) *The Ashgate Research Companion to Moral Panics*, Suerry: Ashgate. doi:10.4324/9781315613307.

Kurzweil, R. (2005) *The Singularity Is Near: When Humans Transcend Biology*, New York: Viking.

Lawson, S. (2011a) "Articulation, Antagonism, and Intercalation in Western Military Imaginaries," *Security Dialogue*, 42, 1: 39–56. doi:10.1177/0967010610393775.

————. (2011b) "Cold War Military Systems Science and the Emergence of a Nonlinear View of War in the US Military," *Cold War History*, 11, 3: 421–40. doi:10.1080/146827 45.2010.494302.

————. (2013) "The US Military's Social Media Civil War: Technology as Antagonism in Discourses of Information-Age Conflict," *Cambridge Review of International Affairs*, 27, 2: 226–45. doi:10.1080/09557571.2012.734787.

————. (2014) *Nonlinear Science and Warfare: Chaos, Complexity, and the U.S. Military in the Information Age*, London: Routledge. doi:10.4324/9780203766446.

LeMay, G.C.E. and Smith, M.G.D.O. (1968) *America is in Danger*, New York: Funk and Wagnalls.

Lenoir, T. (2000) "All But War Is Simulation: The Military-Entertainment Complex," *Configurations*, 8, 3: 289–335. doi:10.1353/con.2000.0022.

————. (2002) "Fashioning the Military Entertainment Complex," *Correspondence: An International Review of Culture and Society*, 10, Winter: 14–16.

————. (2003) "Programming Theaters of War: Gamemakers as Soldiers," in Latham, R. (ed) *Bombs and Bandwidth: The Emerging Relationship Between Information Technology and Security*, New York: New Press.

Liddell Hart, B.H. (1925) *Paris; Or the Future of War*, New York: E.P. Dutton & Company.

Lind, W.S. (1979a) "Military Doctrine, Force Structure, and the Defense Decision-Making Process," *Air University Review*, May–June.

————. (1979b) "A Resposne," *Air University Review*, November–December.

————. (1985) *Maneuver Warfare Handbook*, Boulder, CO: Westview Press.

Lyotard, J.-F. (1984) *The Postmodern Condition: A Report on Knowledge*, Minneapolis: University of Minnesota Press.

MacDougall, R. (2006) "The Wire Devils: Pulp Thrillers, the Telephone, and Action at a Distance in the Wiring of a Nation," *American Quarterly*, 58: 715–41. doi:10.1353/aq.2006.0062.

Marwick, A.E. (2008) "To Catch a Predator? The Myspace Moral Panic," *First Monday*, 13, 6. doi:10.5210/fm.v13i6.2152.

Marx, L. (1994) "The Idea of 'Technology' and Postmodern Pessimism," in Smith, M.R. and Marx, L. (eds) *Does Technology Drive History? The Dilemma of Technological Determinism*, Cambridge, MA: MIT Press, pp. 237–58.

———. (1997) "Technology: The Emergence of a Dangerous Concept," *Social Research*, 64, 3: 965–88.

Mattelart, A. (2000) *Networking the World, 1794–2000*, Minneapolis, MN: University of Minnesota Press.

Metcalfe, J. (2016) "Mapping Hundreds of Power Disruptions Caused by Squirrels," *CityLab*, 12 January 2016. Online. Available: <www.citylab.com/design/2016/01/map-squirrel-cyberattack-power-outage-blackout-terrorism/423648/> (accessed 12 January 2016).

Mitchell, G.R. (2006) "Team B Intelligence Coups," *Quarterly Journal of Speech*, 92, 2: 144–73. doi:10.1080/00335630600817993.

Mohr, C. (1984) "On Military: Debating Frills and Fundamentals," *The New York Times*, 3 April 1984, A28.

Nye, D.E. (1994) *American Technological Sublime*, Cambridge, MA: MIT Press.

———. (2010) *When the Lights Went Out: A History of Blackouts in America*, Cambridge, MA: MIT Press. doi:10.7551/mitpress/8252.003.

Oreskes, N. and Conway, E.M. (2010) *Merchants of Doubt: How a Handful of Scientists Obscured the Truth on Issues From Tobacco Smoke to Global Warming*, New York: Bloomsbury Press.

Ricks, T.E. (2009a) "The Coindinistas," *Foreign Policy*, 30 November 2009a. Online. Available: <www.foreignpolicy.com/articles/2009/11/30/the_coindinistas> (accessed 30 November 2009a).

———. (2009b) *The Gamble: General David Petraeus and the American Military Adventure in Iraq, 2006–2008*, New York: Penguin Press.

Rochlin, G.I. (1998) *Trapped in the Net: The Unanticipated Consequences of Computerization*, Princeton, NJ: Princeton University Press. doi:10.1515/9781400822263.

Romjue, J.L. (1984) *From Active Defense to AirLand Battle: The Development of Army Doctrine, 1973–1982*, Fort Monroe, VA: Historical Office: United States Army Training and Doctrine Command.

Sherman, W.C. (1926) *Air Warfare*, New York: The Ronald Press Company. doi:10.21236/ada421698.

Simon, L. (2004) *Dark Light: Electricity and Anxiety From the Telegraph to the X-Ray*, Orlando, FL: Harcourt.

Simpson, C. (1994) *Science of Coercion: Communication Research and Psychological Warfare, 1945–1960*, New York: Oxford University Press.

Standage, T. (1998) *The Victorian Internet: The Remarkable Story of the Telegraph and the Nineteenth Century*, New York: Walker and Co.

Starry, G.D.A. (1978) "A Tactical Evolution," *Military Review*, August: 2–11.

Stiehm, J. (1996) *It's Our Military, Too!: Women and the U.S. Military*, Philadelphia: Temple University Press.

Sullivan, G.G.R. and Coroalles, L.C.A.M. (1995) *The Army in the Information Age*, Carlisle Barracks, PA: Strategic Studies Institute, U.S. Army War College.

Sullivan, G.G.R. and Dubik, C.J.M. (1993) *Land Warfare in the 21st Century*, Carlisle Barracks, PA: Strategic Studies Institute, U.S. Army War College.

Thierer, A. (2013) "Technopanics, Threat Inflation, and the Danger of an Information Technology Precautionary Principle," *Minnesota Journal of Law Science & Technology*, 14: 309. doi:10.2139/ssrn.2012494.

Thrall, A.T. and Cramer, J.K. (eds) (2009) *American Foreign Policy and the Politics of Fear: Threat Inflation Since 9/11*, London: Routledge. doi:10.4324/9780203879092.

Toffler, A. (1970) *Future Shock*, New York: Bantam.

———. (1984) *The Third Wave*, New York: Bantam.

———. (1993) *War and Anti-War: Survival at the Dawn of the 21st Century*, Boston: Little Brown & Co.

Tomes, R.R. (2007) *U.S. Defense Strategy From Vietnam to Operation Iraqi Freedom: Military Innovation and the New American Way of War, 1973–2003*, London: Routledge. doi:10.4324/9780203968413.

Travers, T.H.E. (1975) "Future Warfare: H.G. Wells and British Military Theory, 1895–1916," in Bond, B. and Roy, I. (eds) *War and Society: A Yearbook of Military History*, New York: Holmes & Meier Publishers, Inc., pp. 67–87.

United States Marine Corps. (1989) *FMFM 1: Warfighting*, Quantico, VA: United States Marine Corps.

Wall, D.S. (2008) "Cybercrime and the Culture of Fear: Social Science Fiction(s) and the Production of Knowledge About Cybercrime," *Information, Communication & Society*, 11, 6: 861–84. doi:10.1080/13691180802007788.

Wiener, N. (1948) *Cybernetics*, New York: J. Wiley. doi:10.1037/13140-000.

———. (1950) *The Human Use of Human Beings; Cybernetics and Society*, Boston: Houghton Mifflin.

Winner, L. (1977) *Autonomous Technology: Technics-Out-of-control as a Theme in Political Thought*, Cambridge, MA: MIT Press.

4 Panic, paralysis, and social collapse

The exaggerated fears of cyber-doom

Introduction

In the last two chapters, we have examined the particular forms that cyber-doom rhetoric often takes in the U.S. debate about cybersecurity, as well as the historical context for the kinds of fears that this rhetoric expresses. We have seen that these fears are not new, but rather, are the most recent manifestation of long-standing fears of what might happen to societies and economies when large, opaque technological systems fail or get out of control. Those fears often include predictions of panic and paralysis, followed by social, economic, or even civilizational collapse. But is that what has really happened in the past when technological systems failed due to natural or human causes? In the pages that follow, we will address this question with the goal of assessing whether these fears, as applied to cybersecurity, are warranted.

As demonstrated in the last two chapters, natural disasters, strategic bombardment (including with the use of nuclear weapons), and large-scale terrorist attacks have all served as key analogies and metaphors in cyber-doom rhetoric. They have shaped our pattern of expectation for what might happen in the event of a cyberattack. But rarely do scholars and policymakers look to see the actual effects that these kinds of incidents have caused historically.

This is representative of another key problem with the dominance of cyber-doom rhetoric, which is that it implies that cyber threats are so new, so unprecedented, that there is no base of existing knowledge to guide our thinking. Instead, we resort to imagination and speculation, not only about the future, but also about the past, for which we actually do have knowledge that could guide us, but which we too often ignore. We see examples of this in discussions of cyber conflict and the laws of war, which too often seek to throw out existing knowledge and norms in the face of supposedly unprecedented technological change (Lawson, 2012; Dunlap Jr (ret), 2011). We see it in discussions of "cyber power" in international relations, which is too often uninformed by knowledge of power relations developed in political science and philosophy (Stevens, 2016). Finally, we see it in the persistent use of plainly inappropriate analogies, like nuclear weapons and warfare, and resistance to developing other frameworks to guide our thinking (Lawson, 2012; Sanger, 2016; Rid and Buchanan, 2018).

But we do have a base of knowledge from which to draw when thinking about the possible impacts of cyber conflict. Historians have documented what happens when infrastructures fail due to intentional attack, natural disaster, or accidents. Additionally, we are not without examples of cyberattacks large and small to examine for insight. In short, our thinking about how realistic the fears of cyber-doom are, as well as the possible futures of cyber conflict, can and should be guided by appropriate historical and sociological research, as well as knowledge of cyber conflict as it actually exists today and in our recent past.

This chapter argues that when we do that, we find that cyber-doom as it is often imagined is unlikely. Indeed, as we will see, strategic bombing, nuclear weapons, natural disasters, and terrorist attacks like the one on September 11, 2001, did not cause the kinds of panic and paralysis, social, economic, and civilizational collapse often contemplated as possible impacts from cyberattacks. Cyber conflict as actually experienced so far falls far short of cyber-doom and is also unlikely to cause these kinds of effects in the future.

In the next section, we will examine the work of historians who have documented what has actually happened in the past when the lights have gone out as a result of natural disaster, accident, or intentional attack. We will also gain insight from the work of sociologists who have documented how communities tend to respond in the face of natural or man-made disasters of various kinds. Next, we will turn to what we already know from our experience of cyber conflict today by looking at the impacts of the most prominent examples of cyber "war."

History and sociology of infrastructure failures

As documented in the previous chapter, theorists of industrial, mechanized warfare at the start of the twentieth century argued that the unique vulnerabilities resulting from society's newfound dependence on interlocking webs of production, transportation, and communication systems could be exploited to cause almost instantaneous chaos, panic, and paralysis in a society (Konvitz, 1990; Biddle, 2002). Of course, even today, planning for disasters and future military conflicts alike, including planning for future conflicts in cyberspace, often relies upon hypothetical scenarios that begin with the same assumptions about infrastructural and societal fragility found in early twentieth-century theories of strategic bombardment.

Some have criticized what they see as a reliance in many cases upon hypothetical scenarios over empirical data (Glenn, 2005; Dynes, 2006; Graham and Thrift, 2007: 9–10; Ranum, 2009; Stiennon, 2009). But, there exists a body of historical and sociological data upon which we can draw, which casts serious doubt upon the assumptions underlying cyber-doom scenarios. Just as neither telegraph "wire devils" nor nefarious Internet hackers were the cause of the economic troubles of 1929 or 2008, so too did the predictions of quick victory from the air miss their mark. Work by scholars in various fields of research, including the history of technology, military history, and disaster sociology, has shown time and again that both infrastructures and societies are more resilient than often assumed by policymakers.

Interwar assumptions about the fragility of interdependent industrial societies and their vulnerability to aerial attack proved to be inaccurate. Both the technological infrastructures and social systems of modern cities proved to be more resilient than military planners had assumed. Historian Joseph Konvitz (1990) has noted, "More cities were destroyed during World War II than in any other conflict in history. Yet the cities didn't die." Some critical infrastructure systems like power grids even seem to have improved during the war. Historian David Nye (2010: 48) reports that the United Kingdom, Germany, and Italy all "increased electricity generation." In fact, most wartime blackouts were self-inflicted in an attempt to throw off the aerial invaders, an effort that usually did not fool the enemy or prevent him from dropping his bombs (Nye, 2010: 65).

Similarly, social systems proved more resilient than predicted. The postwar U.S. Strategic Bombing Survey, as well as U.K. studies of the reaction of British citizens to German bombing, all concluded that though aerial bombardment led to almost unspeakable levels of pain and destruction, "antisocial and looting behaviors . . . [were] not a serious problem in and after massive air bombings" (Quarantelli, 2008: 882) and that "little chaos occurred" (Clarke, 2002: 22). Even in extreme cases, such as the atomic bombing of Hiroshima, social systems proved remarkably resilient. A pioneering researcher in the field of disaster sociology describes that

> [W]ithin minutes [of the Hiroshima blast] survivors engaged in search and rescue, helped one another in whatever ways they could, and withdrew in controlled flight from burning areas. Within a day, apart from the planning undertaken by the government and military organizations that partly survived, other groups partially restored electric power to some areas, a steel company with 20 percent of workers attending began operations again, employees of the 12 banks in Hiroshima assembled in the Hiroshima branch in the city and began making payments, and trolley lines leading into the city were completely cleared with partial traffic restored the following day.
>
> (Quarantelli, 2008: 899)

Even in the most extreme cases of aerial attack, people neither panicked, nor were they paralyzed. Strategic bombardment alone was not able to exploit infrastructure vulnerability and fragility to destroy the will to resist of those who were targeted from the air (Freedman, 2005: 168; Nye, 2010: 43; Clodfelter, 2010).

In the aftermath of the war, it became clear that theories about the possible effects of aerial attack had suffered from a number of flaws, including a technological determinist mindset, a lack of empirical evidence, and even willfully ignoring evidence that should have called into question assumptions about the interdependence and fragility of both technological and social systems. In the first case, Konvitz (1990) has argued, "The strategists' fundamental error all along had been [giving] technology too much credit, and responsibility, for making cities work – and [giving] people too little." In his study of U.S. bombardment of Germany, Clodfelter (Clodfelter, 2010) concluded that the will of a nation is determined by multiple factors, both social and technical, and that it therefore takes more than targeting any one technological system or social group to break an enemy's will to resist. Similarly, Konvitz (1990) concluded, "Immense levels of physical destruction simply did not lead to proportional or greater levels of social and economic disorganization."

Next, theories of strategic bombardment either suffered from a lack of supporting evidence or even ignored contradictory evidence. Lawrence Freedman (2005: 168) has lamented that interwar theories of strategic bombardment were implemented despite the fact that they lacked specifics about how results would be achieved or empirical evidence about whether those results were achievable at all. Military planners were not able to point to real-world examples of the kind of social or technological collapse that they claimed would result from aerial attack. But they were not deterred by this lack of empirical evidence. Instead, they maintained that "the fact that infrastructure systems had not failed . . . is no proof that they are not susceptible to failure" and instead "emphasized how air raids *could* exploit the same kind of collapse that *might* come in peace" (emphasis added; Konvitz, 1990). Airpower theorists were not even deterred by seemingly contradictory evidence. Instead, such evidence was either ignored or explained away. For example, during the 1930s, New York City suffered a series of blackouts that demonstrated that the social disruption caused by the sudden lack of power was not severe. In response, airpower theorists deployed the kind of counter-factual thinking we saw in cyber-doom rhetoric in Chapter 2. They argued that the results would have been different had the blackouts been the result of intentional attack (Konvitz, 1990). But the airpower theorists missed the mark in that prediction too. Instead of leading to panic or paralysis, intentional aerial bombardment of civilians "angered them and increased their resolution" (Nye, 2010: 43; Freedman, 2005: 170).

The social reaction to strategic bombardment is just one example of how efforts both to carry out, but also to defend against, such attacks often led to results that were the opposite of what was predicted or intended. One study of the mental health effects among victims of strategic bombing found that excessive precautionary measures taken in an attempt to prevent the panic and paralysis predicted by theorists did more to "weaken society's natural bonds and, in turn, create anxious and avoidant [sic] behavior" than did the actually bombing (Jones et al., 2006: 57). Similarly, in cases of intentional, self-inflicted blackouts, fear of what might happen to society were the power grid to fail led to a self-inflicted lack of power that not only did not have the desired military effect but may also have been an example of excessive, counter-productive precaution that contributed to widespread anxiety (Nye, 2010: 65).

The flawed assumptions and predictions of the airpower theorists had political and military impacts as well. By creating fear of a massive, German reprisal from the air, the promise of mass destruction from the air that military planners had offered civilian policymakers factored heavily into the British decision not to enter the war sooner to stop Hitler's aggression (Biddle, 2002: 2). Once the war began, the failure of the theorists' vision did not lead them to give up on the dream of strategic bombardment, but only to "heavier, less discriminate bombing." As noted in the previous chapter, historian Tami Davis Biddle (2002: 9) has argued, "The result was nothing less than a form of aerial Armageddon played out over the skies of Germany and Japan."

Even though the vision of the airpower theorists had been proven false, assumptions about the fragility of modern societies did not disappear when the war ended. The first use of atomic weapons at the close of the war combined with the beginning of the Cold War nuclear standoff with the Soviet Union kept the old assumptions alive. Surely, U.S. military planners believed, atomic weapons could achieve what strategic bombardment with conventional weapons had not. Thus, fearing "that the American civilian population might collapse in the face of atomic attack," the U.S. military began to support empirical research into the ways that people respond in disaster situations (Quarantelli, 2008: 896). Ironically, the results of that research have consistently called into question the military assumptions that were the original motivation for funding the study of disasters.

Disaster researchers have worked to define more clearly the concepts at the heart of dominant assumptions about how people respond to disaster. Official planning documents, news, and entertainment media alike often assume that in crisis situations people will either be paralyzed or panicked. On the one hand, paralysis can involve "passivity and inaction" in the face of an overwhelming situation (Quarantelli, 2008: 887). This reaction is dangerous because individuals, groups, and entire societies are not able to help themselves and others if they are paralyzed by fear. On the opposite extreme, psychologists and sociologists have defined panic as a heightened level of fear and emotion by an individual or group leading to a degradation of rational thinking and decision-making, a breakdown of social cohesion, and ultimately to injudicious and counterproductive actions that bring more harm or threat of harm (Clarke, 2002: 21; Clarke and Chess, 2009: 998–9; Jones et al., 2006: 58). In short, both paralysis and panic are maladaptive responses to fear, one an under-reaction, the other an overreaction.

Perhaps surprisingly, empirical research has shown repeatedly that "contrary to . . . popular portrayals" by media and officials, "group panic is relatively rare" (Clarke, 2002: 21). Even specific antisocial behaviors such as looting, which is often believed to be a widespread problem in the wake of most disasters, has proven to be "unusual in the typical natural and technological disasters that afflict modern, Western-type societies" (Quarantelli, 2008: 883). Instead of panic or paralysis, "decades of disaster research shows that people behave rationally in the face of danger" (Dynes, 2006). Empirical research has shown that "survivors usually quickly moved to do what could be done in the situation," that their "behavior is adaptive" rather than maladaptive, and that such behavior usually includes "widespread altruism that leads to free and massive giving and sharing of goods and services." The survivors themselves "are truly the first responders in disasters" (Quarantelli, 2008: 885–8). Instead of panic or paralysis leading to social collapse, existing social bonds and norms of behavior are the key assets to effective response, in part because they serve to constrain tendencies towards paralysis, panic, antisocial, or other types of maladaptive behavior (Johnson, 1987: 180).

These results have been confirmed by studying various disasters, both large and small, intentional and accidental, technological and natural, including large-scale blackouts, hurricanes, and terrorist attacks. For example, attacks upon the electrical grid are often featured prominently in cyber-doom scenarios. But historically, just what has happened when the power has gone out? As mentioned earlier, a series of blackouts in New York City in the 1930s indicated that people did not panic and society did not collapse at the loss of electrical power (Konvitz, 1990). That pattern continued through the remainder of the last century, where "terror, panic, death, and destruction were not the result" of power outages. Instead, as Nye (2010: 182–3) has shown, "people came together [and] helped one another," just as they do in most disaster situations.

In August 2003, many initially worried that the two-day blackout that affected 50 million people in the United States and Canada was the result of a terrorist attack. Even after it was determined that it was not, some wondered what might happen if such a blackout were to be the result of intentional attack. One commentator hypothesized that an intentional "outage would surely thwart emergency responders and health-care providers. It's a scenario with disastrous implications" (McCafferty, 2004). But the actual evidence from the actual blackout does not indicate that there was panic, chaos, or "disastrous implications." While the economic costs of the blackout were estimated between 4 and 10 billion dollars (Minkel, 2008; Council, 2004), the human and social consequences were quite minor. Few if any deaths are attributed to the blackout.[1] A sociologist who conducted impromptu field research of New York City residents' responses to the incident reported that there was no panic or paralysis, no spike in crime or antisocial behavior, but instead, a sense of solidarity, a concern to help others and keep things running as normally as possible, and even a sense of excitement and playfulness at times (Yuill, 2004). For example, though the sudden loss of traffic lights did lead to congestion, he notes that the situation was mitigated by

> people spontaneously taking on traffic control responsibilities. Within minutes, most crossing points and junctions were staffed by local citizens directing and controlling traffic. . . . All of this happened without the assistance of the normal control culture; the police were notably absent for long periods of the blackout.
>
> (Yuill, 2004)

James Lewis (Lewis, 2006) of the Center for Strategic and International Studies has observed, "The widespread blackout did not degrade U.S. military capabilities, did not damage the economy, and caused neither casualties nor terror."

Despite the fact that historical and sociological evidence has shown that "people are irked but not terrified at the prospect" of power loss (Nye, 2010: 191), and, therefore, that intentional attacks on the power grid are "not likely to cause the same type of immediate fear and emotion" as a conventional attack (Stohl, 2007), scenarios in which the loss of power leads to panic, chaos, and social collapse persist because of the persistence of a technological determinist mindset among officials, the media, and the general public. Nye has observed that most reports that are written about blackouts after the fact focus on technical reasons for failures and technical or bureaucratic changes to avoid such failures in the future. They "establish many facts but provide little insight into the social meaning or the historical significance of blackouts" (Nye, 2010: 4). Not surprisingly, most of the policy response to the 2003 blackout focused on technical or regulatory changes (Minkel, 2008). What gets overlooked in these accounts and the types of policy responses they encourage is the human capacity for "adaptation and improvisation in the face of crisis" (Nye, 2010: 195).

As mentioned earlier, some have argued that a so-called cyber 9/11 could approximate or even exceed the impacts of the terrorist attacks of September 11, 2001. Others, including the sponsors of cybersecurity legislation, as well as a former White House cybersecurity czar, have spoken of a possible "cyber Katrina" (Epstein, 2009). But, in both of those cases, people generally responded in the ways that they have in other disasters, without panic, paralysis, or social collapse. Disaster sociologist Lee Clarke has noted that on 9/11, "people did not become hysterical but instead created a successful evacuation" (Clarke, 2002: 23). That evacuation of Lower Manhattan, which involved nearly half a million people, "was a self-organized volunteer process that could probably never have been planned on a government official's clipboard" (Glenn, 2005). At the level of the national economy, the Congressional Research Service concluded, "The loss of lives and property on 9/11 was not large enough to have had a measurable effect on the productive capacity of the United States" (Makinen, 2002). A 2010 report by the Center for Risk and Economic Analysis of Terrorism Events showed that the overall economic impacts of the 9/11 attacks were even lower than initially estimated, indicating that the U.S. economy is more resilient in the face of disaster and intentional attack than commonly assumed (*NBC Los Angeles*, 2010). At the geopolitical level, if the goal of the terrorists was to drive the United States from the Middle East, then the 9/11 attacks backfired. Just as World War II aerial bombardment often served to strengthen rather than weaken the will to resist among targeted populations, Freedman (2005: 169) has observed, "The response [to 9/11] was not to encourage the United States to abandon any involvement with the conflicts of the Muslim world but to draw them further in."

Finally, analysis of Hurricane Katrina by disaster sociologists has shown that while there was some looting and antisocial behavior in the immediate aftermath of the disaster, people generally did not panic and Katrina did not result in the kind of social chaos and collapse often implied in media coverage of the event. Quarantelli (2008: 888–9) reports that "pro-social and very functional behavior dwarfed on a very large scale the antisocial behavior that also emerged. [. . .] [This] prevented the New Orleans area from a collapse into total social disorganization." Like the attacks of 9/11, though the economic impacts of Katrina were severe, especially for those areas in the Gulf Coast that were immediately affected, Katrina did not have the effect of collapsing the entire U.S. economy. And while some suggested that U.S. military operations in Iraq slowed the National Guard's response to Katrina (Gonzales, 2005), there was no indication that military response to Katrina had a negative effect upon U.S. military operations overseas or overall military readiness.

The empirical evidence provided to us from historians and sociologists about the impacts of infrastructure disruption, both intentional and accidental, as well as peoples' collective response to disasters of various types, calls into question the kinds of predictions one finds in the cyber-doom scenarios. If the mass destruction of entire cities from the air via conventional and atomic weapons generally failed to deliver the panic, paralysis, technological and social collapse, and loss of will that was intended, it seems unlikely that cyberattack would be able to achieve these results. It also seems unlikely that a "cyber 9/11" or a "cyber Katrina" would result in the loss of life and physical destruction seen in the real 9/11 and Katrina. And if the real 9/11 and Katrina did not result in social or economic collapse, nor in a degradation of military readiness or national will, then it seems unlikely that their "cyber" analogues would achieve these results.

The wages of cyber "war"

As we saw in the last section, we are not without knowledge to guide our understanding of what happens when infrastructures fail, either through accident or intentional attack. Researchers have demonstrated repeatedly that there is often a gap between our imaginations and the realities of those events' impacts. However, none of the kinds of events we examined in the last section were cyberattacks. Some might argue that the impacts would have been different were these events caused by a cyberattack. Of course, this is the same logic that early twentieth-century airpower theorists used when they argued that electrical grid failures due to strategic bombing would have a different effect on the population than electrical failures due to accidents. That turned out to be incorrect. Nonetheless, though the past may be prologue, it is not a perfect predictor of the future. So, in this section we will look more specifically at several of the largest cyberattacks to date that have been deployed as evidence in support of cyber-doom rhetoric. What we will see is that even the most severe cyberattacks, what some have even called acts of cyber "war," fall far short of the impacts imagined in cyber-doom rhetoric.

Estonia, 2007

One of the oft-cited examples of cyber "war" took place in April and May 2007 in the small, Baltic country of Estonia, a former Soviet Republic. In this incident, government and private networks in Estonia came under massive distributed denial of service (DDOS) attack after the Estonian government announced that it would move a Soviet-era war memorial, the Bronze Soldier of Tallinn, from the center of the capital city, Tallinn, to a cemetery on the outskirts of town. Russia objected to moving the memorial and this sparked a row between the two nations. Thus, when DDOS attacks hit the websites of the Estonian parliament and government ministries, as well as private banks, newspapers, and broadcasters, many assumed that the attack was sponsored or encouraged by the Russian government.

Though attribution has never been proven conclusively, assumptions of Russian involvement remain dominant to this day (Blank, 2008; Evron, 2008).

During and after the incident, various commentators provided their analysis of what happened and the implications. Of the attack, *Wired* reported, "Hackers Take Down the Most Wired Country in Europe" (Davis, 2007). As recently as 2016, the attack was called, "The cyberattack that changed the world" (O'Neill, 2016). Stephen Blank of the U.S. Army's Strategic Studies Institute said the attack represented "the advent of . . . new forms of military operations" and Europe's "first information war" (Blank, 2008: 227). Fred Kaplan echoed this sentiment, writing that the DDOS attacks on Estonia (and Georgia too, discussed later) mark "the dawn of a new era in cyber warfare – the fulfillment of a decade's worth of studies [and] simulations" (Kaplan, 2016: 165). Another observer used the Estonia case to push back against what he saw as complacency about cyber threats. He asked, "What would happen if tomorrow the Internet ceased to function?" He continued,

> To most critics, and particularly state officials and policy makers, the possibility that the Internet could one day suddenly disappear is no more than a mere speculation, a highly improbable concept. On May 2007, the events that took place in Tallinn, the capital of Estonia, proved everyone wrong. On that day, Estonia fell victim to the first-ever, real Internet war.
>
> (Evron, 2008: 121)

As we saw in Chapter 2, Amit Yoran, former head of the Department of Homeland Security's National Cyber Security Division, likened the Estonia incident to a "cyber-9/11" (Singel, 2009). The most hyperbolic response, however, was reserved for the speaker of the Estonian parliament, who said, "When I look at a nuclear explosion, and the explosion that happened in our country in May, I see the same thing" (Poulsen, 2007).

The impacts of the 2007 Estonia incident did not approximate what happened on 9/11, and certainly not nuclear warfare. Neither the immediate, tactical impacts, nor the longer-term, strategic impacts, come close to matching what is imagined in the cyber-doom scenarios that Estonia is often used to bolster. In terms of the immediate impacts, Estonia was not, in fact, "taken down," nor even knocked off the Internet. A scientist at the NATO Cooperative Cyber Defence Centre of Excellence, which was established in Tallinn in response to the 2007 incident, wrote that the immediate impacts were "minimal" or "nonexistent," and that "no critical services were permanently affected" (Ottis, 2010: 72). James Lewis wrote in 2009 that neither the Estonia, nor the Georgia DDOS attacks were "acts of war" because "in neither case were there casualties, loss of territory, destruction, or serious disruption of critical services" (Lewis, 2009: 3). Computer security expert Bruce Schneier's assessment, presented at the 2010 Conference on Cyber Security in Tallinn, echoed these sentiments. He told the audience, "It's kind of like an invading army coming into your country and then getting in line at the motor vehicles bureau so you can't renew your drivers license. It's not really what I think of when I think of war" (Schneier, 2010). Similarly, Bill Blunden concluded, "The whole affair ended with a whimper" (Blunden and Cheung, 2014: loc 5026).

It is important to remember, as Valeriano and Maness remind us, that much of the virtual damage that was inflicted was the result of Estonia's response to the attack, not to the attack itself. They write,

> Although the lives of many Estonians were inconvenienced, life went back to normal after the cyber dispute had ceased. In fact, many of the worst impacts of the disputes were self-imposed by Estonia, in order to protect the state from further incursions. [. . .] It is not at all clear that the incident had much of an impact beyond self-inflicted wounds by the Estonians. They choose [sic] to cut off Internet access in response to the actions.
>
> (Valeriano and Maness, 2015: 144–5)

Of course, this sounds quite similar to the self-inflicted electrical blackouts as a largely counterproductive tactic for mitigating the effects of strategic bombing during World War II.

But what about the longer-term, strategic effects of the Estonia DDOS attacks? Here too, it seems that the incident did not have much effect, or at least, not the desired effects for the attacker. Even Kaplan, who claimed that Estonia marked a new era in cyber war, admits, "The Estonian operation was a stab at political coercion, though in that sense it failed: in the end, the statue of the Red Army soldier was moved from the center of Tallinn to a military graveyard on the town's outskirts" (Kaplan, 2016: 165). Ottis notes, "The biggest impact of the cyber attacks has been political," in that it awoke Estonia and NATO to the potential of cyber conflict, leading to development of cyber defense policies for both, as well as providing a "healthy boost" to the establishment of the Cooperative Cyber Defence Centre of Excellence (Ottis, 2010: 72).

In fact, instead of coercing or intimidating Estonia, the attacks seem to have had the opposite effect. Valeriano and Maness note that "the reaction by Estonia was not to move closer to the Russian sphere of influence, but to fully commit to the West" (Valeriano and Maness, 2015: 112). Similarly, instead of crippling or even degrading Estonia's cyber systems in the long term, Estonia is now known as a leader in cyber technologies and cybersecurity. Despite the attacks, Estonia is "more wired than ever. The attacks failed to dent the citizens' confidence in the e-revolution that has brought so much wealth and convenience." Mansfield concludes, therefore, "The attack on Estonia provides an object lesson that what doesn't kill you makes you stronger" (Mansfield-Devine, 2012: 17, 20).

Jason Healey made a similar point in his keynote address to the 2016 Conference on Cyber Conflict in Tallinn, saying,

> [W]hile the Estonians lost the battle, they won the war. The Estonians refused to be coerced and are now renowned for their cybersecurity excellence, while NATO was warned of the dangers of cyber conflict, even building a new NATO cyber centre of excellence in Tallinn. Russia was thereafter known

as a cyber bully. When expressed in terms of longer-term national security outcomes, it is clear they won both operationally and strategically.

(Healey, 2016: 37)

In the end, Valeriano and Maness conclude from all of this that the Estonia conflict is not representative of a future of devastating cyber warfare. Instead, they argue that it

> represents the banality and the trivial nature of cyber conflict as it is inflated in popular perceptions. The conflict is more about what could have occurred rather than what actually happened, given that the Estonian preemptive termination of Internet links was a protective measure.
>
> (Valeriano and Maness, 2015: 142)

In short, it was and is more about imagination and counterfactuals than about the realities of cyber conflict. Instead of the realities of this incident shaping our understanding of cyber conflict, as is so often the case in the cybersecurity debate, our imaginations have distorted our view of reality.

Georgia, 2008

A little over a year after the Estonia incident, the world witnessed yet another series of cyberattacks against the small state of Georgia, another former Soviet republic. These attacks were also widely believed to be the work of Russians. Once again, the pattern of initial reaction to these attacks, followed by more sober assessments of them, was very similar to what we witnessed in the Estonia incident. That is, what started as hysterics about an ominous new age of "cyber war" was followed by later assessments demonstrating the limited effect of these attacks in the context of the overall conflict of which they were a part.

The cyberattacks in question came in the context of an armed struggle between Russia and Georgia in August 2008. In turn, that armed struggle was part of a long-standing, historical feud between the two over the disputed provinces of South Ossetia and Abkhazia. On August 7, 2008, the Georgian army attacked separatist forces in the province of South Ossetia. The following day, Russian military forces responded by invading South Ossetia in defense of the separatists. The Georgian military was no match for the much larger Russian force, leading to quick victory for Russia (Rid, 2013: 28–9).

Alongside the physical conflict, there was also a series of cyberattacks on Georgia that caught the world's attention, in part because the cyberattacks actually began in July 2008, preceding the physical invasion by several weeks (Rid, 2013: 28–9). This suggested to many observers that the cyberattackers had advance notice of the upcoming invasion and, by extension, must have been working with, or encouraged to some degree by, the Russian government. The cyberattacks mainly took two forms, distributed denial of service (DDOS) attacks and website defacements. In the first case, government and private websites were targets, including the Georgian Parliament website, news media sites, and the largest national bank in Georgia. In the second, various websites were defaced, including Georgia's National Bank, the Ministry of Foreign Affairs, and the site for the Georgian President, Mikheil Saakashvili (Bumgarner and Borg, 2009: 4–5; Rid, 2013: 28–9; Valeriano and Maness, 2015: 3–4).

We see in some responses to the Georgia attacks the very tactics of cyber-doom rhetoric we encountered in Chapter 2. First, some exaggerated the immediate impacts of the attacks. For example, Korns and Kastenberg said that Georgia was "cyber locked" and "barely able to communicate on the Internet" (Korns and Kastenberg, 2008: 60) and Bumgarner and Borg argued that the cyberattacks significantly impeded Georgia's ability to respond to the Russian invasion (Bumgarner and Borg, 2009: 6). As we will see shortly, these are most likely overstatements of the attacks' effects during the conflict.

Unsurprisingly, some saw in the Georgia case the fulfillment of years of warnings about impending cyber-doom. As noted earlier, Fred Kaplan pointed to the Georgia attacks as "the dawn of a new era in cyber warfare –the fulfillment of a decade's worth of studies [and] simulations" (Kaplan, 2016: 165). Others have taken the opposite view, not seeing in Georgia the fulfillment of cyber-doom, but rather, a portent of things to come. For example, a 2009 McAfee report asked whether the age of cyber war had finally arrived and whether, more specifically, the Georgia case was a harbinger of things to come. The report quoted Scott Borg, director of the U.S. Cyber Consequences Unit think tank, who said, "So far this technique has been used in denial-of-service and other similar attacks. In the future it will be used to organize people to commit more devastating attacks" (Kurtz et al., 2009: 6). In another report, Borg and co-author Bumgarner claimed that the Georgia attacks somehow demonstrate the vulnerability of oil and gas infrastructure to cyberattacks (Bumgarner and Borg, 2009: 7), despite the fact that they provided no evidence that cyber attackers had anything to do with physical targeting of oil and gas infrastructure by Russian troops or local militants, and even admitted that "cyber attackers refrained from carrying out the sorts of attacks that would have done lasting physical damage to the Georgian critical infrastructure" (Bumgarner and Borg, 2009: 4–5). Similarly, Clarke and Knake saw in *what did not happen* in Georgia, evidence of *what might yet happen* in the future. They wrote,

> [T]hose operations [in Georgia] do not begin to reveal what the Russian military and intelligence agencies could do if they were truly on the attack in cyberspace. The Russians, in fact, showed considerable restraint in the use of their cyber weapons in the Estonian and Georgian episodes. The Russians are probably saving their best cyber weapons for when they really need them, in a conflict in which NATO and the United States are involved.
>
> (Clarke and Knake, 2010: 28)

Finally, the most disturbing response to the Georgia case has included efforts to reconcile "cyber war" and the law of war by seeking to redefine "war" in general to include all of the activities lumped together under the term "cyber war." For example, after concluding that the cyberattacks against Georgia did not constitute "armed attack" under current definitions of that term in the law of war, a report from the NATO Cooperative Cyber Defence Centre of Excellence (CCDCOE) concluded that "new approaches to traditional LOAC [law of armed conflict] principles need to be developed." It advocated that the advent of "new bloodless types of warfare" means that "the definition of an 'attack' should not be strictly connected with established meanings of death, injury, damage and destruction" (Tikk et al., 2008: 30). This could open the door to armed response to bloodless cyber "attacks" like DDOS and web defacement.

Again, as with the Estonia case, the reality of the cyberattacks on Georgia and their effects turns out to be tamer than these assessments would admit. First, after more than a decade or warnings about impending cyber-doom, Valeriano and Maness note that the Georgia attacks were actually evidence of considerable restraint in the use of cyberattack, which their own research shows is part of a larger pattern of cyber restraint among state rivals (Valeriano and Maness, 2015: 16, 95). Similarly, though they used the Georgia case to warn of what might happen in the future, Bumgarner and Borg also noted, "The fact that physically destructive cyberattacks were not carried out against Georgian critical infrastructure industries suggests that someone on the Russian side was exercising considerable restraint" (Bumgarner and Borg, 2009: 5). Thus, the Georgia case was not the fulfillment of more than a decade of cyber-doom predictions. In fact, it seems to have been just the opposite.

Second, the cyberattacks on Georgia seem not to have had much impact on the overall conflict itself. Again, Valeriano and Maness note that the cyber conflict was not a cause the physical conflict, which, as noted earlier, was rooted in a much deeper dispute between Russia and Georgia. Nor was the cyber campaign integral to the Russian campaign or victory (Valeriano and Maness, 2015: 102, 148). There is no doubt that the Russians would have defeated Georgian forces without the cyberattacks. Thomas Rid agrees, writing,

> The effects of the episode were again rather minor. Despite the warlike rhetoric of the international press, the Georgian government, and anonymous hackers, the attacks were not violent. [. . .] The entire affair had little effect beyond making a number of Georgian government websites temporarily inaccessible. The attack was also only minimally instrumental.
>
> (Rid, 2013: 29–30)

Even the Russians seem to have shared this assessment of their Georgian cyber operations. Keir Giles reviewed Russian military and security professionals' own assessment of their country's performance in the war. He noted the existence of "a perception in parts of the Russian Armed Forces that the 'information war' against Georgia had been lost" and that "the common perception among those writing in open sources about the information aspect of the conflict was that the performance of the Russian military in this area badly needed to improve" (Giles, 2011: 1–2).

Finally, those already predisposed to believe that the future of war will be about cyber offense and defense paid close attention to the Georgia conflict primarily for what occurred in cyberspace. But, Peter Singer contrasts the cyber component of the war with the effects wrought by Russian tanks and bombers,

> At the same time, several brigades of Russian tanks crossed into Georgia, and Russian bombers and missiles pummeled the country, causing over 1,300 casualties. The difference in impact was stark. The cyber-attacks alone were not war, but a war was clearly taking place.
>
> (Singer, 2014: 125)

It is perhaps unsurprising, therefore, that the Congressional Research Service report on the implications of Russia-Georgia war for U.S. interests made no mention of the cyberattacks at all (Nichol, 2008).

Iran, 2009

Even some proponents of impending cyber-doom might agree that, in retrospect, the Estonia and Georgia incidents did not live up to the hype. But surely, they would claim, the case of Stuxnet is proof positive that cyber-doom is a real possibility that is only growing more certain by the day. As in the previous two cases, however, more sober assessments of Stuxnet's impacts and lessons paint an altogether different picture. Instead of pointing to the ubiquity, effectiveness, and ease of the kinds of cyber-physical attacks contemplated in cyber-doom rhetoric, Stuxnet was instead an outlier in the world of cyber conflict, one that was both expensive and largely ineffective.

As we saw in Chapter 2, Stuxnet was the name given to a joint, U.S.-Israeli operation carried out against Iranian nuclear facilities in 2009 and 2010, the official name for which was Olympic Games. The operation occurred within the larger context of a geopolitical struggle over the Iranian nuclear program, which Iran claimed was for peaceful purposes, but many others, including the United States and Israel, were convinced was meant for producing weapons. The Olympic Games series of operations was meant to help slow Iranian enrichment and buy time for economic sanctions and diplomacy to work in halting Iranian nuclear activities. Stuxnet was just part of this larger operation, which also included a series of preparatory cyber espionage campaigns meant to help map and pinpoint the vulnerabilities in the uranium enrichment operations at the Natanz facility in Iran. The Stuxnet portion of the operation employed custom malware designed specifically to damage the model of centrifuge used by Iran to enrich uranium (Sanger, 2012a, b; Zetter, 2014). Stuxnet has received a great deal of attention, and for good reason. It remains one of the few examples of cyberattack to cause physical damage and is still one of the most sophisticated pieces of malware yet seen (Valeriano and Maness, 2015: 26; Rid, 2013: 81, 84, 135).

Nonetheless, as with the previous two cases, there was no shortage of hype bordering on hysteria at times in response to Stuxnet. We saw in Chapter 2 that

Stuxnet is used often when the tactic of projection is employed in cyber-doom rhetoric. For example, Stuxnet was deployed as evidence in support of Secretary of Defense Leon Panetta's 2012 warning about an impending cyber Pearl Harbor (Deseret Morning News, 2012; Lieberman et al., 2011). The key rhetorical move here has been to see in Stuxnet itself, or in Iran's cyber response to Stuxnet, evidence of a cyber-doom threat pointed at the United States. As Kaplan noted, the NSA saw its very own Stuxnet operation as "the latest, most dramatic illustration of what agency analysts and directors had been predicting for decades: what we can do to them, they can someday do to us – except that 'someday' was now" (Kaplan, 2016: 213). As we will see, however, Stuxnet was not on par with the cyber Pearl Harbor scenarios the NSA and others had predicted for years, and neither was Iran's response.

Nonetheless, this did not stop politicians, industry experts, and journalists from describing Stuxnet in the most dire terms. For example, an April 2011 article in *Vanity Fair* declared, "Stuxnet is the Hiroshima of cyber-war. That is its true significance, and all the speculation about its target and its source should not blind us to that larger reality. We have crossed a threshold, and there is no turning back" (Gross, 2011). There was no shortage of such commentary calling Stuxnet the cyber equivalent to the dropping of the atomic bomb, calling it a digital missile, and claiming that it heralded a new era in warfare (Lindsay, 2013: 366). At minimum, most media accounts and government statements have portrayed the Stuxnet attack as incredibly successful, having destroyed at least 1,000 centrifuges and delaying the Iranian nuclear program for anywhere from one to five years (Valeriano and Maness, 2015: 95; Barzashka, 2013: 49–50). Indeed, the *New York Times* at one point even claimed that Stuxnet was the "biggest single factor" in delaying Iran's nuclear program (Quoted in Barzashka, 2013: 50).

Like the Estonia and Georgia cases, time and more careful assessment has tempered our understanding of Stuxnet's impacts and lessons. As is too often the case when reporting on incidents of cyberattack, Jon Lindsay argues, "Most accounts of Stuxnet have focused on its unprecedented technical wizardry rather than evaluation of its strategic consequences" (Lindsay, 2013: 368). Indeed, it is hard not to stand up and take notice of Stuxnet. But that is precisely because it is an outlier in a sea of cyberattacks that rank "moderate to low" in the severity of their impacts, with as yet no events leading to "massive damage of critical infrastructure" (Valeriano and Maness, 2015: 94). As Thomas Rid noted in 2013, Stuxnet was "the only known exception" to the fact that "almost all acts of computer-sabotage to date have been non-violent, harming neither machines nor human beings" (Rid, 2013: 135). That is, if we only focus on Stuxnet's technical wizardry, we may be tempted to inflate its immediate effects and its importance as a representative of cyber conflict more broadly. But, in fact, Stuxnet is the exception, not the rule, for cyber conflict and, as we will see, was not particularly effective, which calls into question how common such attacks will be in the future.

Stuxnet appears not to have been successful at meeting its aims. In his analysis of Stuxnet, Jon Lindsay concluded that the attack's impacts on the Iranian enrichment program were "minor and temporary" at best (Lindsay, 2013: 390).

Other analysts have reached largely the same conclusion: Stuxnet was largely a failure in terms of having any significant and lasting impact on the Iranian enrichment program (Blunden and Cheung, 2014: 1073; Valeriano and Maness, 2015: 154). Ironically, Lindsay notes that the need for anonymity, often assumed to be a strength of cyberattack, was an important factor in limiting Stuxnet's effectiveness (Lindsay, 2013: 399). Thus, not only does Stuxnet cast doubt on whether such large-scale attacks can remain anonymous at all, it also casts doubt on whether anonymity is really a strength, as commonly assumed.

What's more, some have argued that Stuxnet may have even backfired, actually helping the Iranians in the long term (Lindsay, 2013: 391; Barzashka, 2013: 48). Lindsay and Barzaska both demonstrate based on reliable, open source data that the number of Iranian centrifuges actually increased over the course of Stuxnet's three waves of attack. Thus, during the time of the attack, Iran actually increased its enrichment of uranium and, thus, its nuclear capabilities. In the longer term, the false impression of Stuxnet's success may have left "Iran to progress quietly," spurred it to "address inefficiencies" in its operations, and slowed diplomatic solutions (Barzashka, 2013).

But beyond its effects on the Iranian nuclear program and efforts to stop it, Stuxnet came with other negative consequences. First, of course, was that the code leaked out "into the wild," which is what allowed private cybersecurity vendors to identify it. Though it did no damage to other systems that it infected, this fact alone was a blow to the idea that cyber weapons are necessarily precise and controllable. Second was that it provoked a response in the form of Iranian cyberattacks against corporate targets in the United States and Saudi Arabia. This included denial of service attacks against a number of U.S. bank websites, as well as an attack on Saudi Aramco that wiped some 30,000 computers. For all the media hype that the Iranian reprisals caused, Blunden and Cheung describe the Iranian reprisals as largely weak and ineffective (Blunden and Cheung, 2014: 5093). Lindsay concurred, arguing that the denial of service attacks "were irritants with little international political consequence or impact on corporate performance" and that Shamoon resembled the "the unsophisticated work of a nationalist hacker" (Lindsay, 2013: 401). Nonetheless, the Iranian response to Stuxnet shows that private firms can end up as collateral damage in a state-level cyber dispute, further undermining the idea that cyberattacks herald a new leap in "precision" warfare. Third, Valeriano and Maness worry that Stuxnet did not just slow diplomatic solutions to the Iranian nuclear problem, but also eroded U.S. moral authority in terms of preventing proliferation of cyber weapons (Valeriano and Maness, 2015: 156). Finally, Barzashka worries that Stuxnet was a waste of cyber capabilities when they were not really needed that will make "future use of cyber-weapons against Iranian nuclear targets more difficult" (Barzashka, 2013: 54).

Ultimately, Stuxnet may indeed herald the future of cyberattacks, but not in the way that many have assumed. As we know from Chapter 2, many have seen in Stuxnet evidence for a future of cyber-doom attacks carried out precisely, anonymously, and on the cheap, leveling the playing field between super powers, rogue states, and non-state actors. But others point to Stuxnet's high cost and mixed results as evidence that Stuxnet-style attacks will remain the exception, rather than the rule, in international conflict. As addressed earlier, Stuxnet was not as anonymous or precise as its designers likely intended and the very quest for anonymity is part of what limited its effectiveness. What's more, Stuxnet was anything but cheap and easy to carry out. A number of scholars have pointed out that Stuxnet required a considerable amount of economic resources and advanced intelligence collection capabilities to pull off, resources and capabilities only available to powerful state actors like the United States and Israel. As a result, they argue that Stuxnet-style attacks will continue to be rare and may be even less likely to be effective (Lindsay, 2013; Rid, 2013; Valeriano and Maness, 2015). Some point to the leaked, U.S. PPD-20 document on cyber warfare as evidence that even the United States has tempered its views of when and how to use cyberattacks as a result of Stuxnet's failures (Kaplan, 2016: 216; Lawson, 2013). Thus, not only do observers like Valeriano and Maness argue that Stuxnet is not evidence of the changing nature of war (Valeriano and Maness, 2015: 153), Thomas Rid goes even further, arguing that Stuxnet was not war at all and that we should stop using war as a frame for thinking about cyber conflict (Rid, 2013: 31, 262–3). At best, he says, Stuxnet was an example of cyber sabotage, not war, and predicts that future cyber sabotage will likely become less violent, not more, as it becomes easier to disrupt systems without destroying them (Rid, 2013: 136). In short, these analysts argue that not only was Stuxnet not cyber-doom, it may even indicate that such scenarios are becoming less, not more, likely.

Ukraine, 2015–2017

Finally, a third former Soviet republic, Ukraine, was the target of several serious cyberattacks between 2015 and 2017. Once again, these attacks have been attributed to Russia. We will recall that Clarke and Knake said that Russia exhibited "considerable restraint" in their use of cyberattacks in Estonia and Georgia and that they were "probably saving their best cyber weapons for when they really need them, in a conflict in which NATO and the United States are involved" (Clarke and Knake, 2010: 28). Professor Timothy Snyder contends that Russia views its conflict in Ukraine as a proxy conflict with the United States and the West (Snyder, 2018). Perhaps this helps to explain the fact that these more recent Russian cyberattacks did indeed have more serious consequences than the incidents in Estonia or Georgia. The attacks against Ukraine included two on the electrical grid that succeeded in shutting down power for a portion of the population for a short period of time in 2015 and again in 2016. They also included the 2017 NotPetya malware attack, which cost organizations worldwide billions of dollars in damages. Though certainly serious, as in previous cases, various observers hyped the impacts of these attacks or used them to warn of impending cyber-doom. None, however, caused the kind of damage or effects contemplated in such scenarios.

On December 23, 2015, cyberattacks on three regional, Ukrainian power companies resulted in power outages that impacted around 225,000 customers for up to six hours (US-CERT, 2016a; Zetter, 2016b). The attacks, which many believe to have been the work of Russia, used spear phishing emails to gain access to, and install malware on, company systems. Once inside, attackers were able to interact with industrial control systems (ICS) directly, shutting down the power (US-CERT, 2016b; Williams, 2017). Attackers also targeted the company with a telephone denial of service (TDOS) attack meant to prevent customers from calling and reporting the power outage (US-CERT, 2016b). Finally, all three companies were targeted with the KillDisk malware that wiped key files from some systems, rendering them inoperable. Though power was restored quickly, up to six months after the attacks, the impacted companies were still "run[ning] under constrained operations" (US-CERT, 2016a).

Almost one year later, on December 17, 2016, another cyberattack targeted the Ukrainian power grid. This time, malware that computer security company Dragos named CrashOverride targeted one Ukrainian power company, causing a brief power outage that impacted fewer people than the 2015 attack (Dragos, 2017a). Other researchers said that the attackers used some of the same malware used in the 2015 attack (Williams, 2017), leading some to speculate that the 2016 attack may also be the work of Russia (Weaver, 2017). Though CrashOverride did not have as much immediate impact as the 2015 attack, experts argued that it was still concerning because it marked an evolution in cyber capabilities against power grids, representing the first tool whose only apparent purpose was to attack electrical grids (Dragos, 2017a). While attackers in 2015 broke in and interacted with ICS directly, CrashOverride was an automated tool that could be used against systems not just in Ukraine, but in other parts of the world as well, and with some modification, even against U.S. power systems (Dragos, 2017a; Weaver, 2017; US-CERT, 2017c).

Then, in June 2017, malware began spreading among computers in Ukraine and, quickly thereafter, around the world. Russian cybersecurity firm Kaspersky Labs dubbed the malware "NotPetya" because of its initial resemblance to the Petya ransomware that had spread worldwide the previous year (Fruhlinger, 2017). Unlike the 2015 and 2016 attacks, however, the U.S. government publicly attributed NotPetya as the work of Russian military intelligence (Nakashima, 2018). Though Petya and NotPetya both encrypt a computer's files, asking the user for payment to get a key to unlock them, in the case of NotPetya, paying the ransom does not actually unlock the files, in effect destroying them instead (Thomson, 2017; Fruhlinger, 2017). The attackers used phishing emails and a malicious update to the Ukrainian MeDoc tax software that leveraged a leaked NSA exploit called EternalBlue to spread their malware (Thomson, 2017; Leyden, 2018). NotPetya ended up impacting computers not just in Ukraine, but around the world, including in Russia (Thomson, 2017; Nakashima, 2018). By 2018, NotPetya was widely acknowledged as "the most costly cyber-attack in history" (Reagan and McCabe, 2018: 1). Andy Greenberg of *Wired* has documented the impact on one of numerous companies affected by NotPetya, the Danish shipping company, Maersk. He noted in 2018 that initial cleanup took ten days, with full cleanup taking two months, all at an estimated cost of $300 million. But several other companies are estimated to have incurred much higher costs than Maersk, including pharmaceutical company Merck, whose damages were reportedly $870 million. A White House assessment of the attack's overall impact put the cost at more than $10 billion (Greenberg, 2018).

Given the centrality of critical infrastructure systems to cyber-doom scenarios over the years, we should expect that these attacks would set off a wave of speculation about the supposed reality of such scenarios. In each case, that is what happened. In response to the 2015 attack, we can see the use of various tactics of cyber-doom rhetoric outlined in Chapter 2. For example, a San Diego newspaper exaggerated the impacts of the attack, claiming that it affected almost three times as many cities as most other reports (Robins, 2016). Several current and former U.S. government officials used the attack to focus not on what did happen, but what could happen. Former director of NSA and CIA Gen. Michael Hayden said the attack was "a harbinger of things to come" (Zetter, 2016a). Director of NSA at that time, ADM Mike Rogers pointed to the attack as evidence that "it is only a matter of the 'when', not the 'if'" the United States would see similar attacks (Gertz, 2016). Officials and experts speculated that the intent of the attacks may have been to weaken Ukrainian citizens' trust in their government or to prevent the proposed nationalization of Ukrainian power companies, a move that Russia opposed. Little evidence was provided to indicate if the cyberattacks actually achieved either of these effects. Nonetheless, Kim Zetter wrote, "Whatever the intent of the blackout . . . (t)his attack was relatively short-lived and benign. The next one might not be" (Zetter, 2016b).

Of course, we now know that "the next one" was even more "short-lived and benign," at least in its immediate impacts. Nonetheless, the 2016 attack was fodder for more cyber-doom rhetoric. Some used it to warn of what did not, but might yet happen. A U.C. Berkeley computer scientist called CrashOverride "a cyber-weapon warhead test," likening it to a "blackout bomb," and warned that it demonstrated that the impacts of an electrical grid attack could be similar to a hurricane or earthquake (Weaver, 2017). Robert Lee of cybersecurity firm Dragos admitted that the immediate impacts of the attack were not significant, but warned of the supposedly heightened "psychological" impacts of future cyber-induced blackouts, as opposed to those caused regularly by weather or animals (Zetter, 2017; Barrett, 2017). Others, like *Wired*'s Andy Greenberg, saw in CrashOverride the fulfillment of longstanding predictions of cyber-doom. "The Cyber-Cassandras said this would happen," he wrote. "Now, in Ukraine, the quintessential cyberwar scenario has come to life. [. . .] They were part of a digital blitzkrieg that has pummeled Ukraine for the past three years – a sustained cyberassault unlike any the world has ever seen" (Greenberg, 2017).

Finally, though widely acknowledged as the most costly cyberattack to date, some still felt the need to deploy cyber-doom rhetoric to discuss the impacts of NotPetya. Again writing for *Wired*, and reminiscent of Estonian officials likening the 2007 DDOS attacks to a nuclear explosion, Greenberg quoted a Ukrainian official saying of NotPetya that "the government was dead" and "it was a massive bombing of all our systems" (Greenberg, 2017). Greenberg shared in these views.

His story made the cover of *Wired*, which proclaimed NotPetya "the code that crashed the world." Inside, Greenberg asserted that "the release of NotPetya was an act of cyberwar by almost any definition" that should remind us that "every barbarian is already at every gate" (Greenberg, 2017).

There is no doubt that these attacks were serious, including for what they may portend about the possible future of other, similar attacks. Nonetheless, as in the cases of Estonia, Georgia, and Stuxnet, that reality, alarming as it may be, does not match the rhetoric of cyber-doom. For example, cyber warfare expert Martin Libicki argued that the impacts of the 2015 cyberattack on the Ukrainian power grid were not much different than a standard blackout (Detsch, 2016). With a population of almost 44 million people,[2] the 225,000 residents impacted by the power outage represent a mere half percent of the population. Robert Lee of Dragos cautioned against the tendency for stories about cyberattacks against critical infrastructure to "get spun out of control by the media" (Lee, 2016). Finally, Pollard and Devost stated flatly that, like the Estonia and Georgia cyberattacks, the 2015 cyberattack on Ukraine's power grid does "not live up to the visions of doom and mass hysteria described in many cyberwar scenarios" (Pollard and Devost, 2016).

Others have reached many of the same conclusions about the 2016 attack. First, it is worth noting that in terms of immediate impacts, the 2016 attack affected even fewer than the half percent of the Ukrainian population impacted in 2015. What's more, though CrashOverride could potentially impact power grids outside of Ukraine, including those in the United States if the malware were modified (US-CERT, 2017c; Dragos, 2017a), the Dragos report said that even a scenario in which it was used against "multiple sites simultaneously" it would "not [be] cataclysmic and would result in hours, potentially a few days, of outages, not weeks or more" (Dragos, 2017a). Dragos' Robert Lee told reporters that in its current form, CrashOverride "doesn't have the capability to start a cascade on the order of the 2003 Northeast U.S. blackout, nor to be easily repurposed to target other industrial control systems like water-treatment plants or gas pipelines" (Poulsen, 2017; see also Zetter, 2017). Additionally, the impacts of the 2016 attack in no way demonstrated that the impacts of a cyberattack could approximate those seen in a hurricane or earthquake, as Weaver argued (Weaver, 2017). In fact, the causality is more likely to run the other way: natural disasters causing the most serious blackouts, not cyber-induced blackouts causing natural disaster-like impacts. A 2017 U.S. National Academy of Sciences report on *Enhancing the Resilience of the Nation's Electricity System* noted that the most serious future threats to the power grid come from climate change-induced natural disasters and listed cyberattacks as a relatively minor concern in comparison (National Academies Press, 2017b: 3–2, 3–22).

Finally, though NotPetya was the most costly cyberattack to date, its impacts still did not approximate the chaos and destruction of cyber-doom scenarios. Regardless of *Wired*'s cover story, it did not, in fact, "crash the world" or even kill the government of Ukraine. What's more, it was not even "an act of cyberwar by almost any definition," as Greenberg claimed (Greenberg, 2017). In 2018, Reagan and McCabe sought to answer the question of whether companies impacted by

NotPetya could invoke the war exclusion in their insurance policies. Their answer: no. The reason was that the economic damage caused by NotPetya, though serious and though caused by a state actor, was not enough to meet the definition of "warlike" activities contemplated in the law of armed conflict on which insurance companies base their own definitions. To meet the definition of warlike, they argued, the impacts of the attack must "go beyond economic losses, even large ones" and include "casualties or wreckage." NotPetya's impacts did not reach that level. Instead, they said that it "bore greater resemblance to a propaganda effort rather than a military action intended for 'coercion or conquest.'" Indeed, in their estimation, "most nation-state hacking still falls into the category of criminal activity" rather than armed attack (Reagan and McCabe, 2018).

Despite recent attention to Ukraine, including the *Wall Street Journal* calling it "cyberwar's hottest front" (Coker and Sonne, 2015) and *Wired*'s Greenberg calling Russian cyberattacks "a sustained cyberassault unlike any the world has ever seen" (Greenberg, 2017), experts argue that Ukraine has been exemplary because of the lack of what we have typically thought of as cyber war in the United States. The 2015 and 2016 grid attacks are significant, they say, precisely because they remind us of what has been largely missing in the Russia-Ukraine conflict, the kinds of cyber war attacks on infrastructure that so many have been predicting and assumed would be part of any such future conflict (Geers, 2015; Detsch, 2016). In fact, what emerges in Russia's operations in Ukraine during this period, as well as against the 2016 U.S. presidential election, is a view of cyber conflict much different than what cyber-doom rhetoric would lead us to expect, a point to which we will return in Chapter 6.

Conclusion

We began this chapter by noting an important contradiction in U.S. public policy discourse about cybersecurity. There is a tendency, on one hand, to frame the expected impacts of cyberattacks in terms of war, terrorism, and natural disaster, while on the other to treat cyber threats as so new and unprecedented as to imply that there is no base of existing knowledge from which to understand our current situation. This chapter responded to this tendency towards "you never knowism" mentioned in Chapter 2. It demonstrated that we do have a solid base of knowledge from historians and sociologists about how societies respond in the kinds of situations most often used to frame the expected impacts of cyberattacks – i.e. strategic bombing, electrical blackouts, catastrophic terrorist attack, or natural disaster. What we learn from these cases is that societies are much more resilient than policymakers and military planners often expect. Even in the most extreme cases, we do not see the panic and paralysis, social, economic, or civilizational collapse often predicted. This has certainly been the case with the most significant incidents of cyber "war" to date, which in no way approach the predicted impacts found in cyber-doom scenarios. Nonetheless, we also saw that expectations of panic and paralysis due to infrastructure failure are themselves quite resilient, seemingly immune to empirical evidence. What's more, even when reality defies their predictions, not only do policymakers and military planners continue to expect and plan for panic and paralysis, but they will often view current events through that lens, seeing in current events the fulfillment of expectations (at least initially) and, in some cases, taking action based on those expectations rather than the reality of the situation. We saw this in the case of early airpower theorists, who continued to expect and plan for panic and paralysis in the event of infrastructure failures, even as real-world examples should have indicated that these predictions were off the mark. More recently, in the case of cyber "war," we see a similar pattern of initially viewing new incidents through the lens of expected cyber-doom, only later to arrive at more sober assessments. This was particularly the case with the incidents from the Ukraine conflict, which, as we saw, resulted in some experts arguing that perhaps we needed to adjust our expectations about the nature and future of international cyber conflict.

We have already begun to catch a glimpse of some potentially negative implications of cyber-doom rhetoric in this chapter, including ignoring relevant history, misapprehension of current events, and responses that are ineffective or even counterproductive. In the next chapter, we will address the negative implications of cyber-doom rhetoric more directly by discussing how fear appeals can backfire, how metaphors and analogies, especially those to war and disaster, can distort our thinking, and some examples of how these negative implications may be impacting our thinking related to cyber threats. We will return, in Chapter 6, to the case of Russian cyber operations, how those have provided the most significant challenge yet to cyber-doom rhetoric, and some of the alternatives that are beginning to emerge as a result.

Notes

1 One report has claimed that as many as eleven deaths can be directly attributed to the blackout (Minkel, 2008).
2 Population data collected from *CIA World Factbook*, https://www.cia.gov/library/publications/the-world-factbook/geos/up.html.

References

Barrett, B. (2017) "Squirrels Keep Menacing the Power Grid. But at Least It's Not the Russians," *Wired*, 18 January 2017. Online. Available: <www.wired.com/2017/01/squirrels-may-beat-power-grid-glad-not-russia/> (accessed 18 January 2017).

Barzashka, I. (2013) "Are Cyber-Weapons Effective?: Assessing Stuxnet's Impact on the Iranian Enrichment Programme," *The RUSI Journal*, 158, 2: 48–56. doi:10.1080/03071847.2013.787735.

Biddle, T.D. (2002) *Rhetoric and Reality in Air Warfare: The Evolution of British and American Ideas About Strategic Bombing, 1914–1945*, Princeton, NJ: Princeton University Press. doi:10.1515/9781400824977.

Blank, S. (2008) "Web War I: Is Europe's First Information War a New Kind of War?," *Comparative Strategy*, 27, 3: 227–47. doi:10.1080/01495930802185312.

Blunden, W. and Cheung, V. (2014) *Behold a Pale Farce: Cyberwar, Threat Inflation, & the Malware-Industrial Complex*, Waterville, OR: Trine Day.

Bumgarner, J. and Borg, S. (2009) "Overview By the US-CCU of the Cyber Campaign Against Georgia in August of 2008," *US-CCU Special Report*, August.

Clarke, L. (2002) "Panic: Myth or Reality?," *Contexts*, 1, 3: 21–6. doi:10.1525/ctx. 2002.1.3.21.

Clarke, L. and Chess, C. (2009) "Elites and Panic: More to Fear Than Fear Itself," *Social Forces*, 87, 2: 993–1014. doi:10.1353/sof.0.0155.

Clarke, R.A. and Knake, R. (2010) *Cyber War: The Next Threat to National Security and What to Do About It*, New York: HarperCollins.

Clodfelter, M. (2010) "Aiming to Break Will: America's World War II Bombing of German Morale and Its Ramifications," *Journal of Strategic Studies*, 33, 3: 401–35. doi:10.1080/01402390903189436.

Coker, M. and Sonne, P. (2015) "Ukraine: Cyberwar's Hottest Front," *Wall Street Journal*, 9 November 2015. Online. Available: <www.wsj.com/article_email/ukraine-cyberwars-hottest-front-1447121671-lMyQjAxMTA2NzIxNTAyMzU4Wj> (accessed 9 November 2015).

Council, E.C.R. (2004) *The Economic Impacts of the August 2003 Blackout*, Washington, DC: Electricity Consumers Resource Council.

Davis, J. (2007) "Hackers Take Down the Most Wired Country in Europe," *Wired*, 21 August 2007. Online. Available: <www.wired.com/2007/08/ff-estonia/> (accessed 21 August 2007).

Deseret Morning News. (2012) "Cyber Security," *Deseret Morning News*, 20 October 2012, LexisNexis.

Detsch, J. (2016) "Did Ukraine Power Grid Hack Give Russia an Edge?," *Christian Science Monitor*, 15 April 2016. Online. Available: <https://m.csmonitor.com/World/Passcode/2016/0415/Did-Ukraine-power-grid-hack-give-Russia-an-edge-video?mc_cid=5a8ec2ad08&mc_eid=de589295fc> (accessed 15 April 2016).

Dragos. (2017a) *Crashoverride: Analysis of the Threat to Electric Grid Operations*, Washington, DC: Dragos.

Dunlap Jr (ret), M.G.C.J. (2011) "Perspectives for Cyber Strategists on Law for Cyberwar," *Strategic Studies Quarterly*, 5, 1: 81–99.

Dynes, R. (2006) "Panic and the Vision of Collective Incompetence," *Natural Hazards Observer*, 31, 2.

Epstein, K. (2009) "Fearing 'Cyber Katrina', Obama Candidate for Cyber Czar Urges a 'FEMA for the Internet'," *Business Week*, 18 February 2009. Online. Available: <www.businessweek.com/the_thread/techbeat/archives/2009/02/fearing_cyber_katrina_obama_candidate_for_cyber_czar_urges_a_fema_for_the_internet.html> (accessed 18 February 2009).

Evron, G. (2008) "Battling Botnets and Online Mobs: Estonia's Defense Efforts During the Internet War," *Georgetown Journal of International Affairs*, 9, Winter/Spring: 121–6.

Freedman, L. (2005) "Strategic Terror and Amateur Psychology," *The Political Quarterly*, 76, 2: 161–70. doi:10.1111/j.1467-923x.2005.00668.x.

Fruhlinger, J. (2017) "Petya Ransomware and NotPetya Malware: What You Need to Know Now," *CSO Online*, 17 October 2017. Online. Available: <www.csoonline.com/article/3233210/ransomware/petya-ransomware-and-notpetya-malware-what-you-need-to-know-now.html> (accessed 17 October 2017).

Geers, K. (ed) (2015) *Cyber War in Perspective: Russian Aggression Against Ukraine*, Tallinn, Estonia: NATO CCDCOE Publications.

Gertz, B. (2016) "CYBERCOM Says Cyberattacks on Infrastructure Coming," *Washington Times*, 9 March 2016. Online. Available: <www.washingtontimes.com/news/2016/mar/9/inside-the-ring-infrastructure-cyberattacks/> (accessed 9 March 2016).

Giles, K. (2011) *Information Troops: A Russian Cyber Command?* Presentation at Cyber Conflict (ICCC), 2011 3rd International Conference on, IEEE.

Glenn, D. (2005) "Disaster Sociologists Study What Went Wrong in the Response to the Hurricanes, But Will Policy Makers Listen?" *The Chronicle of Higher Education*, 29 September 2005. Online. Available: <https://chronicle.com/article/Disaster-Sociologists-Study/120178/> (accessed 29 September 2005).

Gonzales, J. (2005) "Iraq Mess Adds to the Problem," *New York Daily News*, 2005. Online. Available: <www.commondreams.org/views05/0901-25.htm> (accessed 2005).

Graham, S. and Thrift, N. (2007) "Out of Order: Understanding Repair and Maintenance," *Theory, Culture & Society*, 24, 3: 1–25. doi:10.1177/0263276407075954.

Greenberg, A. (2017) "How an Entire Nation Became Russia's Test Lab for Cyberwar," *Wired*, 20 June 2017. Online. Available: <www.wired.com/story/russian-hackers-attack-ukraine/> (accessed 20 June 2017).

———. (2018) *The Untold Story of Notpetya, the Most Devastating Cyberattack in History*, 22 August 2018. Online. Available: <www.wired.com/story/notpetya-cyberattack-ukraine-russia-code-crashed-the-world/> (accessed 22 August 2018).

Gross, M.J. (2011) "A Declaration of Cyber-war," *Vanity Fair*, 1 April 2011. Online. Available: <www.vanityfair.com/news/2011/04/stuxnet-201104> (accessed 1 April 2011).

Healey, J. (2016) "Winning and Losing in Cyberspace," in Pissandis, N., Roigas, H. and Veenendaal, M. (eds) *Proceedings of the 8th International Conference on Cyber Conflict (Cycon)*, IEEE, Tallinn, Estonia, pp. 37–49. doi:10.1109/CYCON.2016.7529425.

Johnson, N.R. (1987) "Panic and the Breakdown of Social Order: Popular Myth, Social Theory, Empirical Evidence," *Sociological Focus*, 20: 171–83. doi:10.1080/00380237.1987.10570950.

Jones, E., et al. (2006) "Public Panic and Morale: Second World War Civilian Responses Re-Examined in the Light of the Current Anti-Terrorist Campaign," *Journal of Risk Research*, 9, 1: 57–73. doi:10.1080/13669870500289005.

Kaplan, F. (2016) *Dark Territory: The Secret History of Cyber War*, New York: Simon & Schuster.

Konvitz, J.W. (1990) "Why Cities Don't Die: The Surprising Lessons of Precision Bombing in World War Ii and Vietnam," *American Heritage Invention & Technology Magazine*, 5, 3: 58–63.

Korns, S.W. and Kastenberg, J.E. (2008) "Georgia's Cyber Left Hook," *Parameters*, Winter: 60–76.

Kurtz, P.B., DeCarlo, D.W. and Simpson, S. (2009) *Virtually Here: The Age of Cyber Warfare*, Santa Clara, CA: McAfee.

Lawson, S. (2012) "Putting the 'War' in Cyberwar: Metaphor, Analogy, and Cybersecurity Discourse in the United States," *First Monday*, 17, 7. doi:10.5210/fm.v17i7.3848.

———. (2013) "Is There a Silver Lining to the President's Cyber War Policy?" *Forbes.com*, 11 June 2013. Online. Available: <www.forbes.com/sites/seanlawson/2013/06/11/is-there-a-silver-lining-to-the-presidents-cyber-war-policy/> (accessed 11 June 2013).

Lee, R.M. (2016) "Potential Sample of Malware From the Ukrainian Cyber Attack Uncovered," *SANS Industrial Control Systems Security Blog*, 1 January 2016. Online. Available: <https://ics.sans.org/blog/2016/01/01/potential-sample-of-malware-from-the-ukrainian-cyber-attack-uncovered> (accessed 1 January 2016).

Lewis, J.A. (2006) "The War on Hype," *San Francisco Chronicle*, 19 February 2006. Online. Available: <https://articles.sfgate.com/2006-02-19/opinion/17283144_1_cyber-attack-pandemic-avian-flu> (accessed 19 February 2006).

———. (2009) "The 'Korean' Cyber Attacks and Their Implications for Cyber Conflict," unpublished manuscript.

Leyden, J. (2018) "A Year After Devastating Notpetya Outbreak, What Have We Learnt? Er, Not a Lot, Says Blackberry Bod," *The Register*, 27 June 2018. Online. Available: <www.theregister.co.uk/2018/06/27/notpetya_anniversary/> (accessed 27 June 2018).

Lieberman, S.J., Collins, S.S. and Carper, S.T. (2011) "Avoiding a Cyber Pearl Harbor," *The Washington Post*, 8 July 2011, A13.

Lindsay, J.R. (2013) "Stuxnet and the Limits of Cyber Warfare," *Security Studies*, 22, 3: 365–404. doi:10.1080/09636412.2013.816122.

Makinen, G. (2002) *The Economic Effects of 9/11: A Retrospective Assessment*, Washington, DC: Congressional Research Service.

Mansfield-Devine, S. (2012) "Estonia: What Doesn't Kill You Makes You Stronger," *Network Security*, July 2012, 13–20. doi:10.1016/s1353-4858(12)70065-x.

McCafferty, D. (2004) "Dark Lessons: Learning From the Blackout of August '03," *Homeland Security Today*, 1 August 2004. Online. Available: <www.hstoday.us/content/view/1177/60/> (accessed 1 August 2004).

Minkel, J.R. (2008) "The 2003 Northeast Blackout—Five Years Later," *Scientific American*, 13 August 2008. Online. Available: <www.scientificamerican.com/article.cfm?id=2003-blackout-five-years-later> (accessed 13 August 2008).

Nakashima, E. (2018) "Russian Military Was Behind 'Notpetya' Cyberattack in Ukraine, CIA Concludes," *The Washington Post*, 2 January 2018. Online. Available: <www.washingtonpost.com/world/national-security/russian-military-was-behind-notpetya-cyber-attack-in-ukraine-cia-concludes/2018/01/12/048d8506-f7ca-11e7-b34a-b85626af34ef_story.html?noredirect=on&utm_term=.5f9a601d4cdb> (accessed 2 January 2018).

National Academies Press. (2017b) *Enhancing the Resilience of the Nation's Electricity System*, Washington, DC: National Academies Press. doi:10.17226/24836.

NBC Los Angeles. (2010) "Study: Economic Impact of 9/11 Was 'Short-lived'," *NBC Los Angeles*, 7 January 2010. Online. Available: <www.nbclosangeles.com/news/business/Study-bin-Ladens-Strategy-Was-Short-Lived.html> (accessed 7 January 2010).

NCCIC/ICS-CERT INCIDENT ALERT, Department of Homeland Security. (2016a) "Alert (iralerth1605601) Cyberattack Against Ukrainian Critical Infrastructure," *NCCIC/ICS-CERT INCIDENT ALERT, Department of Homeland Security*, 2 February 2016a. Online. Available: <https://ics-cert.us-cert.gov/alerts/IR-ALERT-H-16-056-01> (accessed 2 February 2016a).

NCCIC/ICS-CERT INCIDENT ALERT, Department of Homeland Security. (2016b) "Ir-alert-h-16–043–01ap Cyber-attack Against Ukrainian Critical Infrastructure," *NCCIC/ICS-CERT INCIDENT ALERT, Department of Homeland Security*, 7 March 2016b. Online. Available: <https://info.publicintelligence.net/NCCIC-UkrainianPowerAttack.pdf> (accessed 7 March 2016b).

Nichol, J. (2008) *Russia-Georgia Conflict in South Ossetia: Context and Implications for U.S. Interests*, Washington, DC: Congressional Research Service.

Nye, D.E. (2010) *When the Lights Went Out: A History of Blackouts in America*, Cambridge, MA: MIT Press. doi:10.7551/mitpress/8252.001.0001.

Olympia Journal Snowe Press Releases. (2009) "Senator Snowe and Chairman Rockefeller Introduce Comprehensive Cybersecurity Legislation," *Olympia Journal Snowe Press Releases*, 1 April 2009. Online. Available: <https://snowe.senate.gov/public/index.cfm?FuseAction=PressRoom.PressReleases&ContentRecord_id=6306ecb2-802a-23ad-4a08-163f03f287da> (accessed 1 April 2009).

O'Neill, P.H. (2016) "The Cyberattack That Changed the World," *The Daily Dot*, 20 May 2016. Online. Available: <www.dailydot.com/layer8/web-war-cyberattack-russia-estonia/> (accessed 20 May 2016).

Ottis, R. (2010) "The Vulnerability of the Information Society," *futureGOV Asia Pacific*, August–September 2010, 70–2.

Pollard, N.A. and Devost, M.G. (2016) "Is Cyberwar Turning Out to Be Very Different From What We Thought?" *Politico*, 6 August 2016. Online. Available: <www.politico.com/magazine/story/2016/08/is-cyberwar-turning-out-to-be-very-different-from-what-we-thought-214136> (accessed 6 August 2016).

Poulsen, K. (2007) "'Cyberwar' and Estonia's Panic Attack," *Threat Level*, 22 August 2007. Online. Available: <www.wired.com/threatlevel/2007/08/cyber-war-and-e> (accessed 22 August 2007).

———. (2017) "U.S. Power Companies Warned 'Nightmare' Cyber Weapon Already Causing Blackouts," *The Daily Beast*, 12 June 2017. Online. Available: <www.thedailybeast.com/newly-discovered-nightmare-cyber-weapon-is-already-causing-blackouts> (accessed 12 June 2017).

Quarantelli, E.L. (2008) "Conventional Beliefs and Counterintuitive Realities," *Social Research: An International Quarterly*, 75, 3: 873–904.

Ranum, M. (2009) *The Problem With Cyberwar*, Presented at DojoSec Monthly Briefings, March 2009.

Reagan, T. and McCabe, M. (2018) "Notpetya Was Not Cyber 'War'," *Marsh Insights*, August 2018. Online. Available: <www.marsh.com/us/insights/research/notpetya-was-not-cyber-war.html> (accessed August 2018).

Rid, T. (2013) *Cyber War Will Not Take Place*, Oxford: Oxford University Press.

Rid, T. and Buchanan, B. (2018) "Hacking Democracy," *SAIS Review of International Affairs*, 38, 1: 3–16. doi:10.1353/sais.2018.0001.

Robins, G. (2016) "Public Yawns at Threat of Cyber Crime," *San Diego Union-Tribune*, 5 June 2016. Online. Available: <www.sandiegouniontribune.com/news/science/sdut-cyber-attacks-public-2016sep05-story.html> (accessed 5 June 2016).

Sanger, D.E. (2012a) *Confront and Conceal: Obama's Secret Wars and Surprising Use of American Power*, New York: Crown Publishers.

———. (2012b) "Obama Order Sped Up War of Cyberattacks Against Iran," *New York Times*, 1 June 2012b. Online. Available: <www.nytimes.com/2012/06/01/world/middleeast/obama-ordered-wave-of-cyberattacks-against-iran.html> (accessed 1 June 2012b).

———. (2016) *Avoiding Cyber Conflict Among Nations: Lessons From the Nuclear Age, the Cold War and the Drone Wars*, Presentation at International Conference on Cyber Conflict, Tallinn, Estonia. Online. Available: <www.youtube.com/watch?v=qjqLdjN8Ko8>.

Schneier, B. (2010) *Keynote Address*, Presented at Conference on Cyber Conflict, Cooperative Cyber Defence Centre of Excellence, Tallinn, Estonia, 18 June 2010.

Singel, R. (2009) "Is the Hacking Threat to National Security Overblown?" *Threat Level*, 3 June 2009. Online. Available: <www.wired.com/threatlevel/2009/06/cyberthreat> (accessed 3 June 2009).

Singer, P.W. (2014) *Cybersecurity and Cyberwar: What Everyone Needs to Know*, London: Oxford University Press.

Snyder, T. (2018) *The Road to Unfreedom: Russia, Europe, America*, New York: Tim Duggan Books.

Stevens, T. (2016) *Power in and Through Cyberspace*, Presentation. International Conference on Cyber Conflict, Tallinn, Estonia. Online. Available: <www.youtube.com/watch?v=Y14ThzI0o5k>.

Stiennon, R. (2009) "Scenarios Are Silly Syllogisms," *ThreatChaos*, 19 October 2009. Online. Available: <https://threatchaos.com/2009/10/scenarios-are-silly-syllogisms/> (accessed 19 October 2009).

Stohl, M. (2007) "Cyber Terrorism: A Clear and Present Danger, the Sum of All Fears, Breaking Point or Patriot Games?" *Crime, Law and Social Change*, 46, 4–5: 223–38. doi:10.1007/s10611-007-9061-9.

Thomson, I. (2017) "Everything You Need to Know About the Petya, er, Notpetya Nasty Trashing PCs Worldwide," *The Register*, 28 June 2017. Online. Available: <www.theregister.co.uk/2017/06/28/petya_notpetya_ransomware/> (accessed 28 June 2017).

Tikk, E., et al. (2008) *Cyber Attacks Against Georgia: Legal Lessons Identified*, Tallinn, Estonia: NATO Cooperative Cyber Defence Centre of Excellence.

US-CERT. (2017c) "Alert (ta17163a)—Crashoverride Malware," *US-CERT*, 12 June 2017c. Online. Available: <www.us-cert.gov/ncas/alerts/TA17-163A> (accessed 12 June 2017c).

Valeriano, B. and Maness, R.C. (2015) *Cyber War Versus Cyber Realities: Cyber Conflict in the International System*, London: Oxford University Press. doi:10.1093/acprof:oso/9780190204792.001.0001.

Weaver, N. (2017) "A Cyber-weapon Warhead Test," *Lawfare Blog*, 14 June 2017. Online. Available: <https://lawfareblog.com/cyber-weapon-warhead-test> (accessed 14 June 2017).

Williams, B.D. (2017) "How Dangerous (and Innovative) Is the Newly Discovered Power Grid Malware?" *Fifth Domain*, 15 June 2017. Online. Available: <https://fifthdomain.com/2017/06/15/how-dangerous-and-innovative-is-the-newly-discovered-power-grid-malware/> (accessed 15 June 2017).

Yuill, C. (2004) "Emotions After Dark: A Sociological Impression of the 2003 New York Blackout," *Sociological Research Online*, 9, 3. doi:10.5153/sro.918. Online. Available: <www.socresonline.org.uk/9/3/yuill.html>.

Zetter, K. (2014) *Countdown to Zero Day: Stuxnet and the Launch of the World's First Digital Weapon*, New York: Crown Publishers.

———. (2016a) "Everything We Know About Ukraine's Power Plant Hack," *Wired*, 20 January 2016a. Online. Available: <www.wired.com/2016/01/everything-we-know-about-ukraines-power-plant-hack/> (accessed 20 January 2016a).

———. (2016b) "Inside the Cunning, Unprecedented Hack of Ukraine's Power Grid," *Wired*, 3 March 2016b. Online. Available: <www.wired.com/2016/03/inside-cunning-unprecedented-hack-ukraines-power-grid/> (accessed 3 March 2016b).

———. (2017) "The Malware Used Against the Ukrainian Power Grid Is More Dangerous Than Anyone Thought," *Motherboard*, 12 June 2017. Online. Available: <https://motherboard.vice.com/en_us/article/zmeyg8/ukraine-power-grid-malware-crashoverride-industroyer> (accessed 12 June 2017).

5 When fear fails

The dangers of cyber-doom rhetoric

Introduction

Thus far, we have identified cyber-doom rhetoric as a form of fear appeal and explored its various tactics. We have traced the fears expressed in this rhetoric to longstanding beliefs in U.S. history and culture about the presumed role of technology in general, and communication technologies specifically, in the formation and maintenance of societies and economies. In Chapter 4, we saw that fears of socioeconomic collapse induced by technological failure have been misplaced and that even the most severe examples of "cyber war" have not lived up to the effects contemplated in cyber-doom rhetoric. Nonetheless, one might still be tempted to say that cyber-doom rhetoric is, well, just rhetoric. One might say that the use of hyperbole and poor metaphors by politicians and others is to be expected and that, if we are indeed misdiagnosing the cyber threats we face, the language we use to talk about these threats is the least of our concerns.

This chapter takes a different view, arguing that language matters in shaping how we perceive and respond to threats. The fictional scenarios that we indulge and the language that we use can shape our pattern of expectation about possible futures, focusing our attention in certain directions and not others, sometimes with dire implications. As noted at the start of this book, some have gone so far as to claim that a focus on fictional cyber-doom scenarios left the United States looking in the wrong direction when Russia struck in 2016. Of course, the reasons for the U.S. failure to see the Russian cyberattacks coming are very likely multiple. There is good reason to believe, however, that obsession with cyber-doom was at least one of them.

In the pages that follow, this argument will be supported first by examining research by communication scholars who have demonstrated that although we are tempted by the siren song of fear appeals, those messages are prone to failure if not deployed carefully. We then turn our attention to research on the importance of metaphor, analogy, and framing for effectively diagnosing and responding to problems. Research by cognitive scientists and others has shown that language has a powerful effect on how we perceive and respond to the world around us. Various scholars have critiqued the American tendency to deploy war metaphors to frame our understandings of, and responses to, social problems of various kinds.

Disaster sociologists in particular have shown that militarized framings of natural disaster contribute to elite fears of mass panic and social paralysis during disaster situations, which in turn encourages counterproductive responses. Next, we will see that cyber-doom and its associated rhetorical tactics suffer from several cognitive distortions, including metaphorical idolatry, two variations of probability neglect, and externalization of responsibility for our actions. Taken together, these distortions can result in distraction from real threats and encourage costly, ineffective, and sometimes even counterproductive responses.

When fear appeals backfire

The first set of negative consequences for cyber-doom rhetoric comes from the fact that it works, as noted in Chapter 2, as an appeal to fear. Scholars as far back as Aristotle have studied the nature of fear appeals, in particular their ethics and effectiveness. What they have found contradicts much of the received wisdom on the topic. We have probably all heard that appeals to emotion are logical fallacies and might, therefore, conclude that fear appeals are fallacious and perhaps even unethical. At the same time, as we saw in Chapter 2, politicians and advocates of various kinds cannot let go of the temptation to use fear, often believing that if a little bit is good for gaining attention and motivating action, then more must be better (Peters et al., 2014). But often the opposite is true in both cases. Research by scholars in rhetoric, communication, psychology, and sociology demonstrates that fear can at times serve as an ethical and effective way to raise awareness and promote positive behavioral changes in audiences. But this research also shows that such effects are very difficult to achieve. Fear appeals can easily turn out to be unethical and counterproductive. In this section, we review the findings from this work and how it relates to cybersecurity discourse.

For centuries, logicians, rhetoricians, and scholars of argumentation have tended to characterize all fear appeals as fallacious or unethical because of their use of emotion and their resemblance to a class or arguments called *argumentum ad baculum*, or "argument to the stick." In this latter form, fear appeals do not just appeal to one's emotions. Rather, the speaker attempts to gain the assent of the audience by actually threatening them with harm, either directly or by withholding protection if the audience does not do what the speaker asks. In essence, it is the argumentative form of the traditional protection racket. More recently, however, several scholars have argued that fear appeals do not always take such form and are, therefore, not always unethical or fallacious (Pfau, 2007; Walton, 2000).

The question then becomes when and how fear appeals may be used effectively and ethically. Rhetorician Michael Pfau identified a number of guidelines in Aristotle's writings on the subject. To be effective, Aristotle advised, fear appeals must convince the listener that the threat in question is actually harmful to the listener and likely to occur, but also that the listener can do something meaningful to help prevent or control the effects of the threat (Pfau, 2007: 231–3). Similarly, to be an ethical use of fear appeal, Aristotle counseled, the threat in question must be real rather than hypothetical. Additionally, fear should be used as a tool for opening up deliberation about finding appropriate courses of action rather than for manipulating audiences into acquiescing to the speaker's wishes. Thus, if audiences are not convinced that a threat is real, harmful, and likely to occur to them, they may reject the speaker's message. Likewise, even if audiences take the threat seriously, they may still reject the speaker's message if they feel that nothing can be done or that there is no role for them to play in the response. This is why Aristotle warned speakers to beware of deploying "overpowering fears" that he said could cause "flight or resignation and inaction" (Pfau, 2007: 227). While a little fear might be good, there is a threshold beyond which more is not better.

Deploying fear appeal models very similar to the one articulated by Aristotle, social scientists have sought to determine just where that threshold for the effective use of fear might be. Early models included message components related to the severity and probability of threat and the effectiveness of the proposed response (Rogers, 1975; Maddux and Rogers, 1983). More recent work expands upon these models, breaking fear appeal message components into two main groups related to threat and response. The threat components include messages related to the severity and susceptibility of the threat for audiences. The response components include messages related to response efficacy and self-efficacy. Echoing Aristotle, these models posit that effective fear appeals should address the harmfulness of a threat to listeners, their susceptibility to that threat, and the degree to which listeners can do something effective in response (Witte, 1994).

Psychologists and communication scholars have carried out much of the research on fear appeals in the context of health and safety campaigns, such as promoting smoking cessation and wearing of seat belts. This research has generally confirmed Aristotle's advice. Successful fear appeals result in listeners taking meaningful action to control or reduce the threat. Failed fear appeals result in inaction or even counterproductive actions that make the threat worse. Researchers initially hypothesized that the more fear the better in terms of achieving successful outcomes. What they found instead was more in line with Aristotle's centuries-old advice: too much fear can be ineffective or even counterproductive. In particular, recent research has demonstrated that successful fear appeals require balancing the threat and response components of the message. If listeners do not believe that there is something that can be done to respond effectively to the threat and that they are capable to carrying out that action, they will reject the fear appeal, no matter how severe the threat or how susceptible to it they are. In fact, without effective response components of the fear appeal message, listeners might develop fatalistic beliefs leading to actions that make the threat worse, such as a smoker who decides not only to continue smoking but to smoke more because he or she believes the situation is hopeless anyway (Witte, 1994; Peters et al., 2013; Kok et al., 2018).

Fear appeal research has only recently begun to be applied in the areas of cyber and information security. These studies mainly focus on the individual or organizational level and examine questions related to motivating people to adopt better cyber and information security practices such as changing passwords or using anti-virus software. So far, the results largely mirror what one would expect from fear appeals research in health and safety campaigns. While threat perceptions, including severity and probability, are an important contributor to peoples' intention to take action – there must be at least some fear in a fear appeal for it to work (Boss et al., 2015) – nonetheless, perceptions of response efficacy and self-efficacy are also necessary to improving people's intention to act (Pfleeger and Caputo, 2012; Siponen et al., 2014). Fear alone is not enough.

For example, one study of anti-virus usage indicated that while "perceived severity of virus threats . . . [was] found to have no significant relationship with virus protection intention," perceived ability to utilize effective anti-virus tools was significantly related to study participants' intention to engage in such behavior (Lee et al., 2008: 449–50). Thus, the authors concluded that those who promote better information security practices "should not only concentrate their efforts into increasing individuals' awareness of the likelihood of virus attacks, but also conduct interventions aimed at increasing self-efficacy and response efficacy beliefs" (Lee et al., 2008: 445). A similar study found that while perceived severity of threat did have an impact on participants' intention to comply with organizational information security policies, perceived probability did not. However, like other studies, perceived efficacy of responses and self-efficacy in implementing them were both related to intention to act. Interestingly, the use of threats of punishment for not following the policy were significantly related to demotivating participants to act, indicating that a more pure, *ad baculum* argument – i.e. the speaker threatening the audience with harm to gain compliance – was actually counterproductive (Herath and Rao, 2009: 116–17). These findings are beginning to make their way into industry. In August 2019, for example, PricewaterhouseCoopers' principal for cybersecurity told *CSO Online* that people need to be empowered to act, not scared straight with cyber-doom rhetoric. "[S]elling on fear, uncertainty and doubt don't [sic] build support for the security program" (Pratt, 2019).

More recent research indicates that these effects might also be found at a wider population level. In Chapter 2, we encountered the 2014 *National Geographic Channel* docudrama *American Blackout* as an example of how fact and fiction blur in media discourses about cybersecurity. We recall that this hour-long program depicted the anticipated effects of a cyberattack that took down the U.S. power grid for ten days. A 2016 study examined this program through the lens of fear appeals research. It concluded that, as a fear appeal message, *American Blackout* deployed over-the-top severity and susceptibility threat components, encouraging viewers to believe that no one would escape the inevitable devastation of such an attack. However, the program included little if any content that conveyed the message that there is something effective that viewers could do to help prevent or mitigate the threat. Analysis of viewer responses to the show were in line with what one would expect from fear appeals research. Many viewers seemed not to take the dramatic depiction seriously. But among those viewers who did, responses "were more likely to also express a sense of fatalism about the threat. Likewise, few responses indicated that viewers believed that there was something efficacious that either they or the government could do to prevent or respond to the scenario." The study concluded, therefore, that "frightening rhetoric of cyber-doom scenarios . . . can be counterproductive to addressing real cyber attack threats, particularly if such messaging leads people to discount or downplay the potential threat" (Lawson et al., 2016: 75). It is worth recalling from Chapter 2 that many mass media fear appeals, including those that depict cyberdoom scenarios, follow a pattern similar to that found in *American Blackout*.

Argumentation scholar Douglas Walton has written most extensively on the ways that fear appeals can become fallacious or unethical and, in the process, a danger to democracy. Communication contexts such as public policymaking that rely on "persuasion dialogue," "critical discussion," and "deliberation" are particularly vulnerable to the use of fallacious or unethical fear appeals. In these cases, the assumption is that the goal of communication is to explore several sides of an issue; explore the best course of action by weighing reasons and evidence, pros and cons; or resolve conflicting views or opinions. In these cases, the introduction of fear or threat can serve to shut down conversation and/or end criticism. That is, in a deliberation on policymaking or planning, an advocate makes an argument to the effect of, "If you don't let me have my way or acquiesce to my plan, horrible things will happen" (fear) or "If you don't let me have my way, I can't protect you from the horrible things that might happen" (veiled threat) (Walton, 2000: 188–91).

Walton offers the U.S. Department of Transportation's (DOT) campaign to promote the 55 mph speed limit as an example. Even though the original goal was to improve fuel efficiency, DOT later promoted 55 mph as a safety measure despite knowing there was little if any research to back that claim. In this case, he says, the government agency used fear to distract from otherwise weak evidence in support of their claim and as a means of shutting down deliberation and getting their way. This deception, Walton argues, is likely fallacious (Walton, 2000: 193–4).

Though deceptive and likely fallacious, nonetheless, the use of particularly fearful real-world examples or even fictional scenarios is a common tactic for raising a sense of susceptibility to a threat in an audience when supporting evidence is otherwise lacking. In the context of health promotion campaigns, for example, it is worth recalling Atkin's point, which we encountered in Chapter 2:

> It often is important to support persuasive incentives with convincing evidence, particularly to augment the credibility of susceptibility claims. For fear appeals where there is a low level of actual vulnerability, the likelihood of harm can be buttressed by depicting rare but vivid cases rather than underwhelming statistical figures; this tactic may also heighten relevance and comprehensibility.
>
> (Atkin, 2002: 51)

Such tactics can work because people have a tendency to overestimate the likelihood of particularly frightening threats like terrorist attacks. The impacts, however, can be quite negative (Cramer and Thrall, 2009: 6). Sunstein observes that use of frightening, worst-case scenarios can encourage "two opposite problems:

excessive overreaction and utter neglect" (Sunstein, 2007: loc. 52). Overreaction and neglect can both end in allowing "hawkish" policy responses, either through overt support in the case of overreaction or acquiescence in the case of neglect. The danger in both cases is that such responses can ultimately be counterproductive and lead to self-fulfilling prophecies – i.e. making the once unlikely threats more likely (Cramer and Thrall, 2009: 6).

Walton adds that fear appeals can also be unethical and fallacious in political discourse and deliberation when they are used as a tactic of misdirection. Using the example of the Willie Horton campaign advertisement used by George H.W. Bush against Michael Dukakis in the 1988 presidential election, Walton says,

> Where the ad baculum argument could perhaps be seen or criticized as a fallacy . . . is in line with its use to shift the dialogue away from other issues and onto a single emotional issue. [. . .] [B]y focusing so much consideration on one single case, this use of argument from a single example diverted considerable attention and discussion away from other considerations.
>
> (Walton, 2000: 200)

He cautions, therefore, about the negative effects that fear and threat appeals can have in democratic political discourse: "To be sure, the use of force, threats, and fear appeals can be destructive to the democratic process. [. . .] [A]d baculum arguments are indeed a potent obstacle to free democratic political deliberations and open critical discussions of political issues" (Walton, 2000: 199).

Likewise, sociologists have argued that the "promotion of fear and the propagandistic manipulation of information" by politicians and advocates of various kinds have contributed to the emergence of the wider "culture of fear" discussed in Chapter 2 (Furedi, 2006: 32–5). They warn that the result is that we are often distracted by inflated worry about the wrong threats (Glassner, 1999: xix–xx). What's more, these distractions are not based on actual experiences but instead on imagined future harms (Furedi, 2006: ix–x), so much so that "the expectation of apocalypse has become rather banal" (Furedi, 2006: 28–9). In response, Sunstein argues that we see a tendency to react to fear by calling for, or sometimes actually implementing, a precautionary approach to laws and regulations without fully weighing their costs (Sunstein, 2005), a trend that Furedi has also documented (Furedi, 2009). Sociologist David Altheide warns about the long-term political implications when authorities use fear, with the help of mass media, as a tool of propaganda to manipulate public opinion and gain support for the authorities' agendas (Altheide and Michalowski, 1999; Altheide, 2002, 2006). These impacts are concerning on their own. But, we will see later in this chapter that each also reflects biased forms of thinking that undermine our abilities to understand and respond to the threats we face.

Cyber-doom rhetoric falls prey to many of these criticisms of fear appeals. Even though, like the DOT, most experts recognize that the bulk of real cyber threats are much more mundane, many still deploy the various tactics of cyber-doom rhetoric to frame cyber conflict primarily in terms of war and disaster. Cyber-doom rhetoric thus directs our attention to the one, most emotionally charged potential outcome while potentially diverting our attention from more realistic threats. In doing so, fear of the catastrophic impacts hypothesized in cyber-doom scenarios encourages acquiescence to centralized government and military responses. For example, in *American Blackout*, we will recall that assistance seemed only to come from government. There were few if any scenes of people helping one another. Instead, by day three of the crisis, people were depicted as rioting and committing violence against one another. This, in turn, was depicted as necessitating the use of force by military and police, as well as government taking control of all food and water supplies. As we will see later in this chapter, this depiction is not surprising because our discourses of disaster response also tend to promote centralized, even militarized government responses. Other depictions of cyber-doom encountered in Chapter 2, such as *CNN*'s *Cyber Shockwave*, also encouraged centralized, militarized responses.

Outside the fictional world of *American Blackout* and similar scenarios, however, it is important to recall that so far, at least, the United States government's primary efforts to deal with cybersecurity threats are found in agencies of the national security state, including the military, intelligence, and law enforcement. A 2019 report on the history of U.S. Cyber Command attributes these developments at least in part to the influence of Secretary Panetta's 2012 cyber Pearl Harbor warning (Stone, 2019). In short, the overwhelming fear found in cyber-doom rhetoric generally, and doom scenarios like *American Blackout* more specifically, can encourage the closing of deliberation on how best to understand and respond to the full range of cyber threats that we face, as well as fatalism on the part of individuals that promotes acquiescence to centralized, even militarized government responses. But such responses, as we will see later in this chapter, can be counterproductive, another reason why continued reliance on cyber-doom rhetoric should give us pause.

Metaphors as frames

Western thought in the Enlightenment tradition has tended to see metaphor as either frivolous, as mere literary flourish, or at other times as a dangerous tool of deception. Whatever the case, it was said to be a poor substitute for clear, literal language, which was to be the gold standard for truly scientific understanding and description of the world. But over the course of the twentieth century, scholars came to understand that "language, perception, and knowledge are inextricably intertwined" (Ortony, 1979: 2; Schön, 1993: 137). As we have seen throughout the book, this is the case in the articulation of knowledge about security too, where metaphors and analogies often serve as important elements in the articulation of security imaginaries. Cyber-doom rhetoric relies on a number of tactics, including historical metaphors and analogies, use of fiction, hypothetical scenarios, conflation, projection, and more to reinforce the twin, master metaphors of war and disaster used to frame cyber conflict. Thus, metaphorical language in particular, and rhetoric more generally, serve as a window into the dominant systems of meaning that power human understanding and, ultimately, actions (Lakoff and Johnson, 1980: 3, 7). It is worth spending some time, therefore, discussing the ways that metaphors work and how they can lead us astray.

As Lakoff and Johnson explained, metaphor is an essential part of the way that humans make sense of the world. They wrote, "Metaphor is not just a matter of language, that is, of mere words. . . . [H]uman thought processes are largely metaphorical . . . the human conceptual system is metaphorically structured and defined" (Lakoff and Johnson, 1980: 6). This means that "[t]he essence of metaphor is understanding and experiencing one kind of thing in terms of another" (Lakoff and Johnson, 1980: 5). As Donald Schön argues, metaphor "refers both to a certain kind of product – a perspective or frame, a way of looking at things – and to a certain kind of process – a process by which new perspectives on the world come into existence" (Schön, 1993: 137).

What's more, metaphors do not just work individually or in isolation but collectively and systemically. First, they help to structure collective, human knowledge. This is where the use of metaphorical language helps to bridge the gap between individual human cognition and collective understanding and action. Scholars and practitioners alike of law (Nerhot, 1991; Hibbitts, 1994; Weinreb, 2005; Lamond, 2008), the natural sciences (Keller, 1995; Cowan et al., 1999; Wyatt, 2004), foreign policy (Jervis, 1976; Khong, 1992; Saperstein, 1997), and military affairs (Libicki, 1997; Paparone, 2008; Bousquet, 2009; Lawson, 2011b; Goldman and Arquilla, 2014) have all noted the central role of metaphors and analogies to the production of knowledge in these fields.

Schön has written in particular about the central role of what he calls "generative metaphors" in public policy debates. Such debates, he says, are dominated by the "problem-solving perspective" where "problems themselves are generally assumed to be given" and our "task is to find solutions to known problems" (Schön, 1993: 143). However, he argues that in reality, "problems are not given" but instead are "constructed by human beings" using language (Schön, 1993: 144). Metaphors, even when deployed only tacitly, are generative of the broader stories we use to "select for attention a few salient features and relations from what would otherwise be an overwhelmingly complex reality." That is, metaphors and the policy stories they generate are used to frame our understanding of the problems we face (Schön, 1993: 146).

Second, metaphors work together in systems and, therefore, come with "entailments." This means that a root metaphor can bring with it other, related metaphors (Lakoff and Johnson, 1980: 9). "The constellation of notions familiarly associated with" a particular metaphor can transform our perception of the object or phenomenon we seek to understand. We may, as a result, notice aspects of the object or phenomenon we seek to understand that would have gone unnoticed before, or understand existing features and relations in a new way (Schön, 1993: 140–1). In the case of cyber conflict, for example, framing our discussion in terms of "war" then entails related metaphors or "associated commonplaces" (Schön, 1993: 140–1) such as "militaries," "weapons," "law of armed conflict," "deterrence theory," and much more (Lawson, 2012a).

Noting these characteristics, scholars have come to see rhetoric, metaphor in particular, as not merely a tool for understanding and describing the world, but as at least partially constitutive of that world (Lakoff and Johnson, 1980: 145–6; Schön, 1993). Metaphors work not only as cognitive but also normative "structuring devices" (Wyatt, 2004: 245). They shape how we understand the way the world is, but also how it should be and the actions that we take based on these beliefs, including "the writing of legislation, the formation of policy, the design of programs, the diligence of planners, the allocation of funds, the conduct of evaluation" (Schön, 1993: 146). Metaphor-powered policy framing stories make a "normative leap" that presents as "obvious . . . the direction for [a problem's] future transformation" (Schön, 1993: 146–7).

This is particularly true when it comes to understanding things that are new or novel, such as the emergence of a new technology and the threats and opportunities it might pose. Applying the biological metaphor of evolving systems to metaphors themselves, Alan Beyerchen argued that "[a]s evolving things, metaphors . . . can capture the underlying processes of other evolving entities surprisingly well" (Beyerchen, 1997: 76). Thus, the hope is that by allowing us "to see similarity in difference and difference in similarity" (Geary, 2011), at its best metaphor can and should help to provide a balanced view of the new and novel in relation to the old and familiar.

But, of course, metaphor, analogy, and rhetoric generally are not always at their best. While we cannot avoid or get beyond metaphor and rhetoric to absolutely literal and "objective" language, nonetheless we should be cautious and reflective about our use of metaphors because they "carry with them, often covertly and insidiously, natural 'solutions'" (Ortony, 1979: 5–6). Not only can our language limit our vision and understanding of the world, but it can also constrain our possible avenues of action (Lakoff and Johnson, 1980: 10). Indeed, there are numerous ways that metaphor and analogy can go wrong.

Schön observed that the metaphors powering our problem setting stories in policy debates get their "normative force from certain purposes and values, certain normative images, which have long been powerful in our culture" (Schön, 1993: 147). Like discourse more generally, our metaphors are often not our own. What we are able to say, and hence able to see and do, are shaped and constrained by the broader systems of rhetoric and discourse in which we are already enmeshed. Thus, instead of providing fresh insight and illumination on the problems we face, the metaphors we use to frame issues and construct policy narratives can serve instead to reify existing ways of seeing and acting.

This is because we can have a tendency to let our metaphors do our thinking for us. Schön writes that the presence of conflicting stories about a situation can "make it dramatically apparent that we are dealing not with 'reality' but with various ways of making sense of a reality" (Schön, 1993: 148–9). Mongoven has described this "conceptual error of mistaking a mediating metaphor for complete literal reality" as "metaphorical idolatry" and warned that it "can become a dangerous self-fulfilling prophecy" (Mongoven, 2006: 403). Von Ghyczy explained that this occurs when we mistake metaphors for models of reality. The first, he said, are useful for opening up our thinking about new and novel situations. The second are good for providing valid, one-to-one mappings between features in a known and a new domain that can then be used to direct our actions in the new domain. Indeed, sometimes metaphors work best at promoting innovative thinking precisely because of their flaws, the aspects that do not map directly from one situation to another. "It is a problem," therefore, that "a potential metaphor is all too often and all too quickly pressed into service as a model." In these cases, we get the benefits neither of metaphors nor models (von Ghyczy, 2003: 90). In short, instead of an aid to knowledge, metaphor and analogy can end up serving as a poor replacement for it (Lapointe, 2011; Libicki, 1997).

The result can be that we end up less knowledgeable about both phenomena and objects implicated in our metaphor or analogy. That is, when we attempt to understand A in terms of B (e.g. cyber conflict in terms of war or natural disaster), we can end up understanding each of these less well than we did before. This is because while attempting to understand A in terms of B can illuminate certain aspects of A that we may have otherwise missed, it can also cause us to overlook important features of A that do not fit with the new analogy we have made to B. Of course, understanding A in terms of B presumes that we actually understand B to begin with (Schön, 1993: 148). It can be a problem, therefore, that when we set A and B in a relation of equivalence to one another through the use of metaphor, our understanding of B can also be altered. The solidity of that which we had attempted to use to understand some novel, complex, or ambiguous situation can be eroded in the process (Schön, 1993: 140–1). As I will argue later, this is the case with cyber conflict. Not only are war and natural disaster inapt metaphors or analogies, but our very understanding of these phenomena can be dangerously altered in our efforts to use them to understand cyber conflict.

The perils of bad (cyber) language

The danger in framing cyber conflict primarily with metaphors to war and disaster and reinforcing them through the use of doom scenarios is that these tactics tend to promote misdiagnosis of problems, biased ways of thinking, and counterproductive responses. War is exemplary of those metaphors that exert a "normative force from certain purposes and values, certain normative images, which have long been powerful in our culture" (Schön, 1993: 147). Americans have long had a propensity for fighting wars, some more metaphorical than others, including wars on disease, poverty, drugs, and terror (Hardt and Negri, 2004: 13–14). In these wars, the American experience during World War II often serves as the preferred underlying model of war – after all, we worry about a cyber Pearl Harbor not a cyber Tet Offensive – used to frame situations and justify our own actions (Noon, 2004: 342). As we saw in Chapter 4, ideas about disaster have a close relationship to ideas about war in American history. Research into the way populations respond during disaster situations, whether human or natural, emerged out of Cold War military concerns related to civil defense in the event of a nuclear war. It is still common today for officials to rely on martial framings when contemplating the impacts of disasters and necessary responses. But we also saw that the assumptions of rapid social and economic collapse during disasters turned out to be wildly off the mark in almost all cases. In fact, as we saw with the examples of intentional blackouts during WWII, as well as the response to Hurricane Katrina, sometimes officials' responses have been decidedly counterproductive. There is reason for concern, therefore, that the war/disaster framing at the heart of cyber-doom rhetoric may encourage similar outcomes in the area of cyber conflict.

Inflation and distraction

First, the use of war metaphors can have a tendency to inflate potential dangers and, therefore, the sense of urgency to act while at the same time distracting us from legitimate problems that are perhaps less amenable to such martial framings. As David Noon notes, "The rhetoric and iconography of war often represent the only available resource for framing contemporary experience" and have been used repeatedly "to elicit public consent for all sorts of disparate ventures." The result, he writes, has been that since the end of World War II, we have seen a "tendency for policymakers to imagine the nation in the midst of a perpetual war" (Noon, 2004: 342). Of course, some of these were actual shooting wars, but many were not. Nonetheless, martial rhetoric, with its entailments and associated commonplaces, can have a similarly inflationary effect in the merely metaphorical "wars" too. For example, Ann Mongoven has documented the ways that the dominant framing of medicine as a "war against disease" results in "present[ing] medicine as facing perpetual crisis (because illness is never totally defeated)" (Mongoven, 2006: 404). And because the model of war imagined in the "war on disease" is most often that of WWII-style "total war" in search of unconditional victory, we can tend to inflate or misdiagnose the dangers we face (Mongoven, 2006: 406–8). While we focus on waging "a heroic war effort" against some diseases (Mongoven, 2006: 406–8), we also engage in "continuous deflections of [other] important health and social agendas" that "are consigned to a symbolically lesser [non-war] sphere" (Mongoven, 2006: 414, 406–8).

As noted in the introductory chapter, numerous observers have claimed that anticipation of cyber-doom distracted the United States from what was coming in 2016. For example, in 2012, James Lewis of the Center for Strategic and International Studies said that Secretary of Defense Leon Panetta's infamous cyber Pearl Harbor speech marked the official militarization of the U.S. government's response to cybersecurity challenges (Lawson, 2012c). In the wake of Russian cyber-enabled information operations during the 2016 U.S. presidential election, however, some came to see this development as having turned out to be a dangerous distraction. One component of "how Russian cyber power invaded the U.S.," a group of prominent cybersecurity reporters for *The New York Times* said, was focus on cyber Pearl Harbor-like doom scenarios. They wrote,

> But American officials did not imagine that the Russians would dare try those techniques [cyber-enabled influence operations] inside the United States.

They were largely focused on preventing what former Defense Secretary Leon E. Panetta warned was an approaching 'cyber Pearl Harbor' – a shutdown of the power grid or cellphone networks.

(Lipton et al., 2016)

James Lewis was even more blunt, arguing in 2017 that in the wake of Secretary Panetta's speech, "By focusing our defense on critical infrastructure as the target for cyberattack, we have created a cyber Maginot Line that our opponents easily move around" (Lewis, 2017).

This is not the first time that anticipation of imminent cyber-doom may have contributed to distracting from more realistic cyber or non-cyber threats. Even before 2016, some were warning that U.S. attention with respect to cyber threats was misdirected towards cyber-doom and away from the less dramatic "death by a thousand cuts" cyber threats related to crime, espionage, and subversion (Singer, 2014: 70; see also Clapper, 2015).

We can also find examples of distraction with cyber-doom as more dangerous non-cyber threats gathered on the horizon. As early as 1994, for example, futurist Alvin Toffler, a best-selling author whose work was influential in the U.S. national security community at the time, warned that terrorists might collapse the U.S. economy by carrying out a cyberattack against the World Trade Center. Through the rest of the decade, experts and officials continued to warn of impending cyber Pearl Harbor-like attacks while reality continued to defy those predictions. For example, on June 25, 1996, CIA Director John Deutch testified before Congress about the imminent threat of cyberterrorist attacks against the United States. Later that very same day, al-Qa'ida struck Khobar Towers in Saudi Arabia with a truck bomb killing 19 and wounding 498, many of them U.S. service members. As officials continued to warn of cyberattacks, al-Qa'ida struck U.S. embassies in Africa and the USS Cole in Yemen, all in the lead-up to their September 11, 2001, attack on the World Trade Center and the Pentagon. As signals of the impending attacks were being missed, Bush administration officials warned in speeches and testimony of the emerging threat of cyberterrorism. In the wake of the 9/11 attacks, one official even admitted that the administration had been expecting a cyberattack instead of a kinetic attack (Lawson, 2016, 2012b, 2011d; Poulsen, 2003).

One may have expected the shock of 9/11 to correct the situation, but it continued. For example, even though DNI James Clapper rejected the notion of "cyber Armageddon" multiple times in 2015 and 2016, he nonetheless continued to rank cyberattack as a higher threat than terrorism. In the months after these statements, the world watched as ISIS continued the violent expansion of its "caliphate" in the Middle East, as well as carried out terrorist attacks in the United States and Europe. As we saw in Chapter 2, officials appropriated the shock and fear generated by these traditional, kinetic terrorist attacks to warn of impending cyberterrorist attacks (Lawson, 2015c). In response, media coverage shifted for a time in December 2015 to imply that ISIS was the top cyber threat instead of either Russia or China, which had been the focus of coverage to that point (Lawson, 2015d).

Of course, we now know this was occurring at the very moment when Russia was beginning its campaign to interfere in the U.S. presidential election less than a year off.

Metaphorical idolatry

In addition to distracting us from more realistic or dangerous threats, over reliance on war metaphors can bias our thinking in a number of ways. First among these is the "metaphorical idolatry" discussed earlier. Noon made similar observations in his analysis of the use of World War II metaphors during the early years of the "war on terror." Too often, he said, these metaphors came to replace attention to the unique details of the situation, so that their usage was more about repeating our historical memories and interpretations than detailed analysis of current situations in light of the actual facts of historical events (Noon, 2004: 343). Likewise, Mongoven noted that reliance on war metaphors in the medical context can make "examining the environmental, social, and psychological influences on health seem like unaffordable indulgence in the face of such a dangerous enemy" (Mongoven, 2006: 407–8).

We have seen this pattern in our public policy discourse about cyber conflict too. Lapointe reminds us that "cyberspace" is itself a metaphor describing a complex socio-technical network of associations as a sort of geographic "place" that can be inhabited, traversed, and potentially even controlled. This comes with the entailments of talking about cyber ecosystems, battlefields, and domains (Lapointe, 2011). Thomas Rid and Peter McBurney have criticized the flawed entailments that come along with describing cyberattack tools as "weapons," arguing that "even some sophisticated and effective instruments of electronic attack *cannot* be sensibly called a cyber-weapon" (Rid and McBurney, 2012: 6). In the wake of the 2016 Russian cyber operations, Rid was even more committed to this position, writing in 2017, "Most weapons analogies break down even more quickly today than they did four years ago. Serious research and wider public debate would be best served if we all stop using the hackneyed moniker 'cyber weapons' entirely" (Quoted in Biller and Schmitt, 2019: 224). A 2019 legal analysis concurred with Rid's view, faulting poor use of analogies and metaphors and concluding that "cyber capabilities cannot logically be categorized as weapons or means of cyber warfare" (Biller and Schmitt, 2019: 183). Over time, however, these metaphors have come "to seem more literal than metaphoric," with the result that we too often forget "that we are in fact talking metaphorically rather than literally." This, in turn, can "circumscribe our thinking" and hamper "substantive discussion of the way ahead" (Lapointe, 2011). What's more, our thinking is circumscribed even more by the fact that official discourse in the United States has tended to focus narrowly on the technical components of cyberspace, ignoring the social, cultural, and cognitive components (Lawson, 2015a).

Over two decades ago, Martin Libicki noted that the metaphor of cyber "deterrence" was itself rooted in, was an entailment of, "the metaphor that information warfare is indeed warfare" (Libicki, 1997: 7). Deterrence is one of those examples where metaphors that may have started off as aides to creative thinking come to substitute for it as they are taken increasingly literally. In 2010, prominent cybersecurity advocates Richard Clarke and Mike McConnell encouraged the nation to engage in an effort similar to Project Solarium during the early years of the Cold War when U.S. defense intellectuals wrestled with the implications of nuclear weapons and ultimately developed the strategy of deterrence (Clarke and Knake, 2010: x; McConnell, 2010). After almost a decade, Congress established the Cyberspace Solarium Commission in 2019 (Johnson, 2019). In the intervening decade, however, many sought to import the results of the earlier discussion directly, attempting to apply deterrence directly to cyber conflict. A metaphor from another time and very different technology was taken increasingly literally. As Libicki notes, however, "Cyberspace is its own medium with its own rules. [. . .] Thus, deterrence and warfighting tenets established in other media do not necessarily translate reliably into cyberspace" (Libicki, 2009: iii). More recently, Fischerkeller and Harknett have also argued that deterrence is not a viable strategy because of the unique features of cyberspace (Fischerkeller and Harknett, 2017). As Andrew Walworth of *Real Clear Politics* noted after a 2018 interview with the *New York Times*' David Sanger, "One way to help our chances" of avoiding the worst of what cyber conflict may have in store for us "is to not equate" cyber and nuclear weapons. "This is a new technology," he wrote, "requiring new paradigms" (Walworth, 2018). Thus, Libicki and Lapointe both caution that while metaphors can be helpful, we must not let them do our thinking for us, that we must ultimately seek to understand issues on their own terms and in their own specificity (Lapointe, 2011; Libicki, 1997). To do otherwise, Libicki warns, "verges on intellectual abuse" (Libicki, 1997: 6).

Finally, attempts to apply deterrence to cyber conflict can end up simplifying or distorting our understanding of deterrence. Nuclear deterrence was historically more nuanced and varied than many applications to cyber conflict would lead us to believe (Freedman, 1989; Lawson, 2012a). Thus, application of deterrence to cyber conflict risks becoming an example of Schön's warning (discussed earlier) that when we attempt to understand A in terms of B, not only may we overlook specifics about A that are important but otherwise do not fit the metaphorical relationship we have established to B, but that our understanding of B might also be altered in the process, further reducing its utility for effectively understanding A. In framing cyber "weapons" as potential tools for achieving the kinds of strategic paralysis imagined in strategic bombardment and use of nuclear weapons, we attempted to apply a type of deterrence that did not fit the reality of cyber conflict. We forget that Cold War nuclear deterrence did not prevent all war, just nuclear war. Instead, the conflict between the United States and Soviet Union was carried out below a threshold that would trigger nuclear retaliation, at the level of conventional proxy wars and covert political warfare. Similarly, the United States may have been able so far to deter large-scale cyberattacks against critical infrastructure, but in doing so shifted the bulk of cyber conflict below the threshold of armed conflict where traditional deterrence is ineffective. Some have even worried that this misunderstanding and misapplication of deterrence to cyber conflict may have had the opposite of its intended effect, deterring the United States from responding to Russian cyber-enabled information operations rather than deterring the Russians from carrying them out in the first place (Healey, 2018c).

Probability neglect

But the danger in relying on metaphors of war/disaster conveyed through hypothetical doom scenarios goes beyond inflation, distraction, and metaphorical idolatry to encourage distorted views of the probability that the worst possible threats will indeed be realized (Sunstein, 2002). As we saw in Chapter 2, cyber-doom rhetoric can include both deterministic and possibilistic forms of thinking, the first of which overestimates the probability of the most extreme threats while the second sees any level of risk as justifying a preventive or precautionary response, often without sufficient consideration of costs. As we saw in Chapter 3, technological determinism more generally has played an important role in American history. Likewise, deterministic thinking has been common in American discourses about war (Noon, 2004; Hasian, 2005).

In its simplest form, deterministic thinking can take the form of describing the "present as the inevitable outcome of the past" (Noon, 2004: 341–2). But it can also take the form of what Hasian calls "necessitous rhetorics" that present a situation as though we have no choice but to respond and in a particular way (Hasian, 2005). Noon notes that this is, in fact, how WWII analogies often get used by policymakers, who use them to promote the supposed obviousness of certain prescriptions rather than provide clear or accurate descriptions of the problems we face (Noon, 2004: 340, 343). Hasian describes a similar usage of "necessitous rhetorics" in national security and wartime discourses (Hasian, 2005).

These deterministic ways to thinking have, in turn, been prominent for decades in U.S. thinking about the Internet and cyber conflict. American cultural discourses about the Internet have included both utopian and dystopian variants of deterministic thinking that together provide a false dichotomy of the future of cyberspace in society, one promising certain salvation and the other inevitable doom (Schulte, 2013: 7). On one hand, the free flow of information through globally networked ICTs was supposed to topple authoritarian regimes and allow the United States to win bloodless victories in future wars. On the other, we were told that because foreign actors could also harness such capabilities, we must guard against the ever-present threat of an inevitable cyber Pearl Harbor (Rid and Buchanan, 2018: 3). So far, at least, not only have these predictions failed, but they have at times blinded us to the more complex realities of the role of cyberspace in international politics. Authoritarian states have had no problem harnessing a technology that was supposed to be their inevitable undoing (Hughes, 2002: 222; Hasian and Lawson, 2018); network-centric warfare did not deliver quick victory in the "war on terror" (Lawson, 2014); and cyber Pearl Harbor has yet to materialize (Lawson and Middleton, 2019). Instead, perhaps the most consequential cyber operation to date came from an authoritarian regime exploiting social media instead of electrical grids – i.e. Russian cyber-interference in the 2016 U.S. presidential election (Rid and Buchanan, 2018: 3).

In Chapters 2 and 3, we began to see some hints that deterministic ways of thinking and speaking can have negative consequences. This should not come as a surprise. A 2017 systematic review of the psychological literature on "deterministic thinking" identified it as "one of the most important cognitive distortions." "[D]eterministic thinking," the study concluded, "plays a destructive role in individual interactions in family and society" (Sefat et al., 2017: 1). This is because, as Baumeister argues, the probability neglect of deterministic thinking encourages externalization of responsibility for our actions. Because in determinism "everything that happens is the only thing that could possibly happen," the "category of the possible and the category of the actual are exactly the same" with the result that "[t]o a determinist, there is no such thing as actual choice, in the sense of having more than one possible option." Though things might have turned out differently if one had acted differently, to the determinist, "you could not possibly have acted differently" (Baumeister, 2009).

One implication is that we tend to overlook or diminish our own responsibility for the decisions we make and values we enact. This can take the form of blaming others, either human or technological, for our own actions (van der Ploeg, 2003: 91–2; Shields, 2005: 506; Glassner, 1999: xxxiv). At best, projecting onto human or non-human others the causes or consequences of our own actions can contribute to the problems of misdiagnosis and distraction discussed earlier. In these cases, Glassner writes, "bad guys [or technologies] substitute for bad policies" in capturing our attention. We blame dangerous, devious others for our problems instead of identifying the systemic issues or bad policies that are the true culprits. Sometimes we do this to ignore or simplify complex, difficult problems we otherwise have not been able to solve (Glassner, 1999: 6–8). We see an example in claims about the threat that Chinese cyber espionage poses to U.S. economic competitiveness. Though China should certainly be condemned for its widely acknowledged use of cyber intrusions to steal intellectual property, nonetheless, this activity is not the primary cause of lagging U.S. economic competitiveness. Instead, the Chinese "threat" comes from the fact they have adopted the kind of policy for government support for science and technology research that the United States once championed but has increasingly abandoned over the last decade or more (Lawson, 2011d; McPherson, 2019).

Externalization of responsibility can also lead us to lower our ethical standards for our own conduct, allowing us to overlook or excuse our own mistakes or misdeeds. Mongoven notes, for example, that in the context of a total war in search of unconditional victory against disease, we too often overlook side effects and other unintended consequences of medical treatments (Mongoven, 2006: 407–8). Similarly, Noon demonstrates how World War II metaphors and analogies rely on a "rhetoric of prophetic dualism" in which the United States supposedly has no choice but to defend good against evil in a world of perpetual crisis and in which our victory is assured precisely because of the supposed goodness of our cause (Noon, 2004: 358). Finally, Hasian has documented the use of "necessitous rhetorics" in U.S. wartime discourses about the use of torture and military tribunals (Hasian, 2005). In each case, the ends justify the means, even if the cure is worse than the disease, because we supposedly had no choice.

At its worst, the externalization and projection encouraged by deterministic thinking can contribute to self-fulfilling prophecy. Though we may not wish to acknowledge our actions, others certainly will and their responses may very well result in the kinds of situations we had hoped to avoid (Buchanan, 2016). This outcome stems from what is known as the "fundamental attribution error," in which we see our own actions as rational and nonthreatening responses to a given situation while attributing the actions of others to their "character, nature, or deeply held motives." As a result, we might fail to see that our own actions are the cause of the other actor's behavior (Cramer and Thrall, 2009: 5). Jason Healey worries that this is precisely the situation in which the United States finds itself after adopting the concept of "persistent engagement" in 2018:

> We cannot forget that our adversaries are sure they are hitting back, not first. They have their own sense of righteous purpose and the United States is seen as the schoolyard bully. This isn't to make any moral equivalence between U.S. cyber operations and theirs, but there is an escalatory equivalence as each side responds tit-for-tat against the campaigns of the other. Nations will respond very differently to cyber deterrence when they are sure they are hitting back, not hitting first.
>
> (Healey, 2018a)

We can see several examples in the history of U.S. cyber conflict. For example, Fred Kaplan notes that U.S. concern over Chinese cyber capabilities was an outgrowth of China observing and then attempting to emulate the United States (Kaplan, 2016: 224). Similarly, he notes that the Stuxnet and Flame attacks against Iran "spurred the Iranians to create their own cyber war unit." In retaliation, the Iranians used these capabilities against the Saudi Aramco oil company and several U.S. banks (Kaplan, 2016: 213). Even in the case of Russian use of cyber-enabled information operations as part of a supposedly new doctrine of "hybrid warfare," we find hints of a similar kind of feedback loop. Though hybrid warfare is often identified as new and uniquely Russian by U.S. policymakers and commentators, several scholars have argued that Russian officials' use of the term "hybrid warfare" is, in fact, meant to convey their perception of U.S. strategy in the Ukraine and Arab Spring conflicts, in particular Libya and Syria, over the last decade (Kofman and Rojansky, 2015; Adamsky, 2015; Galeotti, 2016). Though these scholars claim that the Russian perception is off the mark, it is not entirely without merit. After all, long before the Russians deployed armies of social media bots and trolls to exploit domestic political rifts, the U.S. national security community was investing in its own capabilities to surveil and manipulate social media (Lawson, 2011c; Webster, 2011; Jarvis, 2011; Nafeez, 2016).

Another variant of distorted views of probability can be found in possibilistic thinking. In this case, even if the worst is not inevitable, its very possibility, no matter how likely, is seen as warranting our response (Clarke, 1999, 2008; Furedi, 2009). Like determinism, possibilistic thinking encourages the conflation of the probable and the merely possible, the conflation of threats and vulnerabilities. As we saw in Chapter 2, it is a conflation that has been particularly prominent in U.S. cybersecurity discourse (Wall, 2008: 867; Dunn Cavelty, 2008: 103). Even the smallest event can be used to focus attention on the worst possible scenario, arguing in effect that anything is possible, that "you never know" what might happen next. Indeed, some have also called such thinking "you never knowism" and warned that it actually encourages taking action based in an absence of evidence, a sort of "faith-based politics" that eschews knowledge in favor of preventive or precautionary action to head off the mere possibility of experiencing a worst-case scenario. The United States' "war on terror" since 9/11 and invasion of Iraq in 2003 are identified as chief examples of the potentially disastrous consequences of such thinking at work in recent history (Furedi, 2009; Handmer and James, 2007; Elmer and Opel, 2008; Lawson, 2011a). It is disconcerting, therefore, to see important voices in the cybersecurity debate, such as former NSA Director and Commander of USCYBERCOM Adm. Mike Rogers, say that "more concerning than known examples of disruptive malware," like the Ukraine cases discussed in Chapter 4, "is what we don't know. The so-called 'known unknown' – malware that remains undetected in our critical infrastructure awaiting instructions to disrupt or even damage in the event of geopolitical escalation" (Rogers and Weinstein, 2019). As we will see shortly, this was also another example of projection, because we would learn only days after Rogers' comments that the United States had been engaged in precisely these kinds of activities against the Russians (Sanger and Perloth, 2019).

In the case of both determinism and possibilistic thinking about technology-related security threats, the proposed actions are themselves often militarized and focused heavily on finding quick technological fixes to complex problems. After all, if threats related to terrorism or cyber conflict are technologically determined, then perhaps a technological solution is in order. As we saw in Chapter 3, early airpower theorists took a technological determinist view of the supposed vulnerabilities of industrial cities and how those could be exploited, often ignoring more nuanced aspects of human relations that turned out to be critical for cities' resilience in the face of aerial bombardment (Konvitz, 1990). We saw a similarly technologically determinist (and possibilistic) framing of threats and supposedly necessary responses in the wake of 9/11. Gaps in the inevitable spread of economic globalization combined with growing technological connectivity were supposedly the root causes of the 9/11 attacks. The same technologies, however, were supposed to allow the United States to act effectively through a combination of electronic surveillance, data mining, and use of military force to prevent the emergence of any similar threats in the future by forcibly "connecting" the supposedly "disconnected" parts of the world (Amoore and Goede, 2005; Chapter 2 in Hasian et al., 2015).

Given this penchant for determinist and possibilistic thinking in U.S. thinking about war and technology, it is perhaps unsurprising that the military and technology have so far dominated U.S. responses to cyber threats. Despite being a complex network of associations that involves more than just technology, the U.S. definition of cyberspace and approach to cyber conflict has been focused on technology (Lawson, 2015a). For example, by its own admission, prior to 2016, U.S. Cyber Command was not focused on the kinds of cyber-enabled information operations seen from Russia in 2016 (Pomerleau, 2018a). When Cyber Command did engage on that front in an operation to counter Russian trolls during the 2018 U.S. midterm elections, the response focused on cutting technical connections of the Russian Internet Research Agency (Barnes, 2019). But the problem of disinformation, foreign and domestic in origin, being used to exploit domestic political rifts is much more than a technical problem. Disrupting the adversary's Internet connection can serve, at best, as a short-term technological fix while leaving root causes untouched.

Ineffective and counterproductive responses

Finally, reliance on metaphors of war and disaster to frame our thinking about public policy problems can encourage counterproductive responses of various kinds. In her examination of the use of war metaphors in medicine and counterterrorism after 9/11, Mongoven observed that this martial rhetoric was associated with responses characterized by overmobilization, centralization, anti-democratic decision-making, and waste. "Because the threat is inflated" as akin to "total war," she writes, then "so is the response" (Mongoven, 2006: 407–8). In both cases, we are willing to bear monetary and political costs beyond what we would normally consider acceptable. This is because "the threat may be conceived as so big that extreme measures are presumed necessary, without any rigorous cost–benefit accounting. [. . .] The threat is described as so big, the times as so extraordinary, that one cannot unduly quibble on idealistic principle" (Mongoven, 2006: 411). On the material side, "the interminable rhetoric of war allows for ever-greater funding allowances" combined with "expansive license to centralized authority" (Mongoven, 2006: 409). On the political side, fear plays a central role in promoting "public acquiescence" (Mongoven, 2006: 409) to top-down, anti-democratic plans in which "ordinary citizens are left with little say as to public priorities or public response" (Mongoven, 2006: 414). She concludes, therefore, that "the structural parallelism between ethical excesses associated with the war against disease and the war on terror warns against unreflective new variations on old metaphorical themes" (Mongoven, 2006: 410).

We may be falling into some of these same traps in our response to cyber threats. A war/disaster framing portends cybersecurity planning dominated by the same "command and control [C2] model" rooted in flawed assumptions of inevitable "panic" and "social collapse" that has increasingly dominated official U.S. disaster planning (Quarantelli, 2008: 897). In disaster planning and responses, the result has been ever more centralized, hierarchical, and bureaucratic disaster responses that increasingly rely upon the military to restore order and official control first and foremost (Quarantelli, 2008: 895–6; Alexander, 2006; Lakoff, 2006).

Likewise, in cyber conflict, a war framing implies the need for military solutions and the establishment of the military's U.S. Cyber Command (USCYBERCOM) has been, to date, the most significant U.S. response to cyber threats.

But the fact remains that most of the malicious activities that get lumped under the term "cyber war" are really acts of crime, espionage, information operations, or political warfare to which it is not at all clear that a military response is most appropriate or effective (Lewis, 2010). Even before the 2016 Russian cyber operations against the U.S. presidential election, military cyber leaders in the United States were coming to this realization based on their experience using cyber capabilities against ISIS. Initial conceptions for how to use cyber capabilities against ISIS tracked along with decades of thinking in the United States that focused on using such capabilities as a substitute for kinetic attacks against enemy command and control (Kaplan, 2016: 14–15, 43). For all the hype in early 2016 about the United States supposedly dropping "cyber bombs" on ISIS command and control, the actual effects of those operations were rather unremarkable and sparked the beginning of a shift in thinking about how cyber capabilities might be used more effectively (Valeriano et al., 2016; Sanger and Schmitt, 2017; Maxey, 2017).

In 2017, Admiral Mike Rogers, then Commander of USCYBERCOM, testified that information operations – i.e. those meant to impact an adversary's perceptions, thinking, or public opinion more broadly – had been largely outside USCYBERCOM's mission set but that this had begun to change during operations against ISIS. Gen. Paul Nakasone, current Commander of USCYBERCOM who led cyber operations against ISIS as head of U.S. Army Cyber Command, made a similar point. In response to this experience, and in the wake of the 2016 Russia operations, Lt. Gen. Stephen Fogarty, commander of Army Cyber Command, speculated that in the future Cyber Command may instead become known as "Information Warfare Operations Command," reflecting a potentially fundamental shift in official understandings of the use of cyber capabilities (Pomerleau, 2018a). In October 2018, Gen. Nakasone said that because most cyber threats fall well below the threshold of armed conflict, DOD must reevaluate its approach. A month later, Lt. Gen. Vincent Stewart, deputy commander of U.S. Cyber Command, was even more blunt, asking, "What if the way we've structured Cyber Command and our thinking about this space, what if it's wrong?" (Pomerleau, 2018b).

A number of critics agree with the underlying concern expressed in Lt. Gen. Stewart's question and say that not only must DOD rethink its approach, but that effective response to the actual cyber threats we face will require more than just the military (Pomerleau, 2018b; Janson, 2018; Seablom and Helms, 2018). So far, at least, recent operations to respond to adversary cyber-enabled information operations resemble the kinds of techniques used prior to 2016. As early as 2010, for example, U.S. operations to counter terrorist recruitment, financing, and propaganda efforts by taking down offending websites made headlines (Nakashima, 2010). Despite such efforts, by 2016, the United States found itself once again in need of dropping supposed "cyber bombs" on ISIS. More recently, and in line with longstanding provisions for USCYBERCOM to provide assistance to the

Department of Homeland Security in the event of a domestic cyberattack (Ackerman, 2010), USCYBERCOM made headlines for responding during the 2018 midterm elections to the kinds of online trolling and propaganda campaigns seen during the 2016 U.S. presidential election that were not commonly framed as examples of cyber "war" or "attack" prior to 2016 (Barnes, 2019). It is not at all clear, however, that such short-term, technical operations will be any more effective in addressing the complex problem of persistent, cyber-enabled information operations by adversaries that exploit domestic political rifts over a long period of time than they were in stopping the rise and spread of ISIS (Janson, 2018; Seablom and Helms, 2018; Healey, 2019).

In addition to perhaps not being the most appropriate or effective form of action, militarized responses to such a broad spectrum of cyber threats are fraught with danger. In the case of disaster response, for example, researchers have demonstrated that militarized framings can contribute to a form of "government paternalism" in which officials panic about the possibility of panic and then take actions that actually exacerbate the situation by not only failing to provide victims with the help they need, but also preventing them from effectively helping themselves (Dynes, 2006; Clarke and Chess, 2009: 999–1001). This phenomenon was on display, for example, in the official response to Hurricane Katrina (Clarke and Chess, 2009: 1003–4). Similarly, we can see examples in the realm of cyber conflict when officials either propose or carry out actions that could or do exacerbate rather than mitigate threats.

First, having a military command with both an offensive and defensive mission as the centerpiece of the U.S. response to cybersecurity threats could undermine the stated U.S. policy of promoting a free and open Internet worldwide by encouraging greater Internet censorship and filtering, as well as more rapid militarization of cyberspace, by adversaries and allies alike (Dunn Cavelty, 2008: 143). The situation is not helped when prominent U.S. politicians advocate for cybersecurity measures reminiscent of those one would expect to see among U.S. adversaries. For example, in 2010, and pointing to China as his inspiration, former Senator Joseph Lieberman proposed a so-called Internet kill switch, which would have given the President the authority to cut U.S. Internet connections to the rest of the world in the event of a large-scale cyberattack (Crowley, 2010; See also McCullagh, 2010). More recently, the Defense Advanced Research Projects Agency (DARPA) has launched a project to develop software that can detect and filter fake news and disinformation from massive amounts of online data, a move that *Bloomberg* framed as the U.S. "unleash[ing the] military to fight fake news, disinformation" (Norman, 2019). These kinds of responses can serve to create a "say-do gap" (Mullen, 2009) that makes it more difficult to criticize others when they seek through the implementation of repressive laws and technological systems to militarize cyberspace and fundamentally fragment the global Internet (Hodge and Ilyushina, 2019; Sherman, 2019).

Second, there is the danger of "blow back." In a highly interconnected world, there is no guarantee that cyber weapons and operations developed or launched by the United States would not result in serious collateral damage to noncombatants or even end up causing harm to the United States itself (Dunn Cavelty, 2008: 143). As it turns out, cyber weapons and operations are not as easy to control as once believed. We should recall, for example, that the world learned about the joint, U.S.-Israeli Stuxnet attack against Iran after the code somehow escaped into the "wild" and was picked up and analyzed by private cybersecurity companies (Zetter, 2014). While the escape of Stuxnet is not known to have caused damage to other systems – it was targeted at very specific features of the Iranian nuclear program not found in other systems – nonetheless, its escape is concerning because it demonstrates the possibility that even the most well-designed and targeted weapons and operations could fall into the wrong hands.

Since Stuxnet, we have seen other cases in which U.S. cyber weapons and operations have caused serious blow back on the United States. In the 2010 terrorist website takedown case mentioned earlier, a U.S. military operation that took down a Jihadist discussion forum caused collateral damage to noncombatant computers and websites and in the process undermined an ongoing U.S. intelligence gathering operation that was using that website as a source (Nakashima, 2010). Additionally, we know from the Snowden leaks that the U.S. government has, at times, and by pointing to cyber threats as a justification, worked to undermine the very technologies and standards that keep the Internet secure (Ball et al., 2013; Gallagher and Greenwald, 2014; Pagliery, 2015; Schneier, 2014). Fred Kaplan also discussed numerous instances of the NSA leaving vulnerabilities un-patched in the belief that they could be used later against U.S. adversaries. The result, he went so far as to argue, was that the government actually made U.S. critical infrastructure less, rather than more secure against foreign cyberattacks (Kaplan, 2016: 139, 280). The implications have become more concerning still since 2016 as exploits developed by the NSA were stolen and used to help enable a series of ransomware attacks against private and government entities around the world, including a number of U.S. cities (Perloth and Shane, 2019; Goodin, 2019; Perloth et al., 2019). Beyond impacting city governments, *CipherBrief* reported, "the estimated cost of damages wrought by the leaking of this cyber weapon has been in the billions of dollars. Foreign intelligence agencies, hacking collectives, and criminal organizations have used Eternal-Blue to target infrastructure including medical, transportation, and financial institutions" (The Cipher Brief, 2019). This tendency may be continuing during the Trump administration, which is reported to be considering a proposed ban on end-to-end encryption – a cornerstone of Internet security – while simultaneously adopting a more aggressive, military doctrine for cyberspace operations (Geller, 2019).

Third, there is the potential for cyberattack to escalate into physical conflict. A decade ago Richard Clarke worried that if the United States launched a cyberattack against a state or non-state actor lacking the capability to respond in kind, that actor might chose to respond with physical attacks (Clarke, 2009). Conversely, there have even been calls for the United States to respond with conventional military force to cyberattacks. In 2011, a U.S. official said that the response to a serious enough hacking attack could come in the form of a "missile down one of your smokestacks" (Gorman and Barnes, 2011). Similarly, retired Air Force

Lt. Gen. Harry Raduege suggested that if a cyberattack were serious enough, "America's response could come in the form of a hellfire missile" (Raduege (Ret), 2011). In other cases, we have seen calls for kinetic responses to cyberattacks that amounted to little more than vandalism (Zetter, 2009; Dunn, 2010). Some have even contemplated a nuclear response. In 2009, a review of U.S. military strategy documents and statements from officials indicated that nuclear response remained on the table as a possible U.S. response to cyberattack (Markoff and Shanker, 2009; Owens et al., 2009).

In general, cyber operations have not been escalatory thus far (Valeriano and Jensen, 2019). There are a few notable exceptions, however, as well as concern that this could change. There are at least two occasions when a state actor responded to cyberattacks with physical force. First, in August 2015, the United States carried out a lethal drone strike against Junaid Hussain, who the United States claimed was an ISIS hacker involved in a hack of the U.S. Central Command website and Twitter account, as well as online recruiting and propaganda activities (Lawson, 2015b). Then, in May 2019, amidst a spate of renewed fighting between Israel and Palestinians in Gaza, the Israeli military carried out an airstrike on a building in Gaza from which it claimed cyberattacks had been launched against Israeli citizens (Fazzini, 2019). Combined with newly aggressive American cyber policies in the Trump administration – often justified with the use of cyber-doom rhetoric – these actions by two powerful militaries who were also involved in the Stuxnet attack against Iran could help to legitimize such actions by other states.

Indeed, there is increasing concern that the risk of further escalations in the severity of cyber conflict, or even from cyber to physical conflict, is increasing as a result of what Kollars and Schneider call the United States' "much more forward-leaning and risk-acceptant cyber strategy" released in 2018 (Kollars and Schneider, 2018). This strategy of "persistent engagement," they explained, was supposed to allow Cyber Command to act more offensively and consistently in cyberspace, focusing its efforts "on degrading adversaries' cyber capabilities instead of threatening attacks on adversary civilian infrastructure" in an attempt "to cripple an opposing nation (which could be highly escalatory)." Nonetheless, they warned that the new strategy could "run the risk of the same sort of escalation the Obama administration sought to avoid" because "we can never be completely sure that more aggressive strategies in cyber space will not spill over to the conventional warfighting domains" (Kollars and Schneider, 2018).

Other scholars have leveled similar concerns. Jason Healey, though agreeing that "persistent engagement" is more appropriate than a strategy of deterrence, warned that the new strategy "ignores many of the risks and how to best address them," including the possibility that it "might not work and significantly increases the chances and consequences of miscalculations and mistakes," all while potentially being "incompatible with the larger U.S. goals of an open and free Internet" (Healey, 2018b). He warned that the United States "must be incredibly cautious escalating the conflict." Especially with respect to Russia, he warned, "Constant contact between opposing cyber forces of nuclear-armed states in a time of turmoil will be destabilizing." He advised, therefore, that "U.S. cyber operations should disrupt operations and command and control of nation-state adversaries, but not go further except in rare circumstances, backed by decisions by political decision-makers" (Healey, 2018a). Though disagreeing with some of Healey's analysis, Max Smeets of Stanford concurred that "this new strategy might be dangerously escalatory" and that the U.S. "has to be careful that it is not creating the opposite effect of what it intends" (Smeets, 2018). Valeriano and Jensen were even more blunt in their critique, writing,

> [R]ecent policy changes and strategy pronouncements by the Trump administration increase the risk of escalation while doing nothing to make cyber operations more effective. These changes revolve around a dangerous myth: offense is an effective and easy way to stop rival states from hacking America. New policies for authorizing preemptive offensive cyber strategies risk crossing a threshold and changing the rules of the game.
>
> (Valeriano and Jensen, 2019: 1)

A shocking report from the *New York Times* in mid-June 2019 seemed to validate the concerns raised by these scholars. The paper reported that United States had been hacking the Russian electrical grid and implanting malware that could be triggered at a later date to cause widespread power outages. The intent of the attacks was to deter Russia from continued hacking of the U.S. grid and interference in upcoming U.S. elections. What's more, the paper reported that Cyber Command had undertaken these activities without fully briefing the President out of fear that he would either disapprove or accidentally reveal the existence of the top-secret operations (Sanger and Perloth, 2019). These revelations set off a wave of criticism and concern from former officials and scholars alike. These operations seemed to go beyond "persistent engagement" and to veer into precisely the territory of "threatening attacks on civilian infrastructure" that Kollars and Schneider had warned "could be highly escalatory" (Kollars and Schneider, 2018) and without the backing of "decisions by political decision-makers" for which Healey had called (Healey, 2018a). Such operations without explicit presidential approval, while likely legal according to law professor Robert Chesney, were still "quite unsettling from the perspective of civil-military relations" (Chesney, 2019). Others pointed out that such operations might not only be ineffective at achieving the intended deterrent effect, but might instead backfire, sparking an escalation in cyberattacks between the two countries the ultimate outcomes of which would be hard to predict but unlikely to be beneficial in any case (Greenberg, 2019; Jensen, 2019).

Conclusion

This chapter argued that the language we use matters for effectively framing and then responding to public policy problems of all types, including cybersecurity. In particular, we explored the ways that two core features of cyber-doom rhetoric – the use of fear appeals and war/disaster metaphors – might lead us astray if not deployed carefully. We saw that fear appeals can easily backfire and that cyber-doom rhetoric exhibits precisely those characteristics that can lead to this outcome. Namely, cyber-doom rhetoric tends to focus our attention on the supposedly over-the-top severity of cyber threats and our susceptibility to them while either ignoring issues of response efficacy or promoting centralized, militarized responses that may not only be ineffective, but perhaps even counterproductive. What's more, we saw that metaphors and analogies can also easily lead us astray and that reliance on war and disaster metaphors in particular can promote a combination of misdiagnosis of problems, overreactions, under-reactions, or counterproductive responses. Together, the use of fear appeal to frame cyber conflict in terms of war and disaster can negatively bias our thinking, encouraging distraction from serious threats less amenable to martial framing, use of metaphors as literal models of reality instead of tools of innovative thinking, and neglect of probability that encourages the externalization of responsibility for our own actions. As we saw in the preceding pages, there is reason for concern that the primary U.S. responses to cyber threats have fallen prey to these negative outcomes. While there is some indication that a shift in thinking is underway, nonetheless, the deeply rooted historical belief in the promise and peril of strategic paralysis by way of infrastructure attacks continues to exert a strong influence over U.S. responses to cyber threats, even in the wake of the 2016 Russian cyber-enabled influence operations against the U.S. presidential election.

In the next chapter, we turn our attention to the U.S. public debate over how to understand and speak about Russia's operations in 2016 and beyond and what they may signal about the future of cyber conflict. We will see that these operations have done the most thus far to undermine the dominance of cyber-doom rhetoric, sparking a serious reassessment of the nature of cyber conflict, including the most appropriate language and analogies for thinking and speaking about that and similar incidents. In the final chapter, we will return to the question of how we might do better in terms of thinking and speaking about cyber conflict more generally.

References

Ackerman, S. (2010) "Doc of the Day: NSA, DHS Trade Players for Net Defense," *Danger Room*, 13 October 2010. Online. Available: <www.wired.com/dangerroom/2010/10/doc-of-the-day-nsa-dhs-trade-players-for-net-defense/> (accessed 13 October 2010).

Adamsky, D. (2015) *Cross-Domain Coercion: The Current Russian Art of Strategy*, Paris: IFRI Security Studies Center.

Alexander, D. (2006) *Symbolic and Practical Interpretations of the Hurricane Katrina Disaster in New Orleans*, Presentation at Understanding Katrina: Perspectives from the Social Sciences, the Forum of the Social Science Research Council, New York: Social Science Research Council. Online. Available: <https://understandingkatrina.ssrc.org/Alexander/>.

Altheide, D.L. (2002) *Creating Fear: News and the Construction of Crisis*, New York: Aldine de Gruyter. doi:10.4324/9780203794494.

———. (2006) *Terrorism and the Politics of Fear*, Lanham, MD: AltaMira Press.

Altheide, D.L. and Michalowski, R.S. (1999) "Fear in the News: A Discourse of Control," *The Sociological Quarterly*, 40, 3: 475–503. doi:10.1111/j.1533-8525.1999.tb01730.x.

Amoore, L. and Goede, M.D. (2005) "Governance, Risk and Dataveillance in the War on Terror," *Crime, Law and Social Change*, 43, 2: 149–73. doi:10.1007/s10611-005-1717-8.

Atkin, C.K. (2002) "Promising Strategies for Media Health Campaigns," in Crano, W.D. and Burgoon, M. (eds) *Mass Media and Drug Prevention: Classic and Contemporary Theories and Research*, Mahwah, NJ: L. Erlbaum, pp. 35–66. doi:10.4324/9781410603845.

Ball, J., Borger, J. and Greenwald, G. (2013) "Revealed: How US and UK Spy Agencies Defeat Internet Privacy and Security," *The Guardian*, 6 September 2013. Online. Available: <www.theguardian.com/world/2013/sep/05/nsa-gchq-encryption-codes-security> (accessed 6 September 2013).

Barnes, J. (2019) "U.S. Cyber Command Bolsters Allied Defenses to Impose Cost on Moscow," *The New York Times*, 8 May 2019. Online. Available: <www.nytimes.com/2019/05/07/us/politics/cyber-command-russian-interference.html?login=email&auth=login-email> (accessed 8 May 2019).

Baumeister, R.F. (2009) "Determinism Is Not Just Causality: Is the Future Already Set in Stone?" *Psychology Today*, 23 June 2009. Online. Available: <www.psychologytoday.com/blog/cultural-animal/200906/determinism-is-not-just-causality> (accessed 23 June 2009).

Beyerchen, A.D. (1997) "Clausewitz, Nonlinearity, and the Importance of Imagery," in Alberts, D.S. and Czerwinski, T.J. (eds) *Complexity, Global Politics, and National Security*, Washington, DC: National Defense University, pp. 70–7.

Biller, J.T. and Schmitt, M.N. (2019) "Classification of Cyber Capabilities and Operations as Weapons, Means, or Methods of Warfare," *International Law Studies*, 95, 179: 179–225.

Boss, S.R., et al. (2015) "What Do Systems Users Have to Fear? Using Fear Appeals to Engender Threats and Fear That Motivate Protective Security Behaviors," *MIS Quarterly*, 39, 4: 837–64. doi:10.25300/misq/2015/39.4.5.

Bousquet, A. (2009) *The Scientific Way of Warfare: Order and Chaos on the Battlefields of Modernity*, New York: Columbia University Press.

Buchanan, B. (2016) *The Cybersecurity Dilemma: Hacking, Trust, and Fear Between Nations*, London: Hurst & Company. doi:10.1093/acprof:oso/9780190665012.001.0001.

Chesney, R. (2019) U.S. Cyber Command and the Russian Grid: Proportional Countermeasures, Statutory Authorities and Presidential Notification," *Lawfare Blog*, 17 June 2019. Online. Available: <www.lawfareblog.com/us-cyber-command-and-russian-grid-proportional-countermeasures-statutory-authorities-and> (accessed 17 June 2019).

The Cipher Brief. (2019) "When Cyberweapons Escape," *The Cipher Brief*, 29 May 2019. Online. Available: <www.thecipherbrief.com/column_article/when-cyberweapons-escape> (accessed 29 May 2019).

Clarke, L. (1999) *Mission Improbable: Using Fantasy Documents to Tame Disasters*, Chicago: University of Chicago Press.

———. (2008) "Possibilistic Thinking: A New Conceptual Tool for Thinking About Extreme Events," *Social Research: An International Quarterly*, 75, 3: 669–90.

Clarke, L. and Chess, C. (2009) "Elites and Panic: More to Fear Than Fear Itself," *Social Forces*, 87, 2: 993–1014. doi:10.1353/sof.0.0155.

Clarke, R.A. (2009) "War From Cyberspace," *The National Interest*, October/November.

Clarke, R.A. and Knake, R. (2010) *Cyber War: The Next Threat to National Security and What to Do About It*, New York: HarperCollins.

Cowan, G.A., Pines, D. and Metzer, D. (eds) (1999) *Complexity: Metaphors, Models, and Reality*, Boulder, CO: Westview Press.

Cramer, J.K. and Thrall, A.T. (2009) "Introduction: Understanding Threat Inflation," in Thrall, A.T. and Cramer, J.K. (eds) *American Foreign Policy and the Politics of Fear: Threat Inflation Since 9/11*, London: Routledge, pp. 1–15. doi:10.4324/9780203879092.

Crowley, C. (2010) "State of the Union With Candy Crowley: Interviews With Senators Lieberman, Murkowski, Feinstein and Lugar," *CNN*, 20 June 2010. Online. Available: <https://edition.cnn.com/TRANSCRIPTS/1006/20/sotu.01.html> (accessed 20 June 2010).

Dunn, J.E. (2010) "North Korea 'Not Responsible' for 4 July Cyberattacks," *Network World*, 6 July 2010. Online. Available: <www.networkworld.com/news/2010/070610-north-korea-not-responsible-for.html> (accessed 6 July 2010).

Dunn Cavelty, M. (2008) *Cyber-Security and Threat Politics: U.S. Efforts to Secure the Information Age*, New York: Routledge. doi:10.4324/9780203937419.

Dynes, R. (2006) "Panic and the Vision of Collective Incompetence," *Natural Hazards Observer*, 31, 2.

Elmer, G. and Opel, A. (2008) *Preempting Dissent: The Politics of an Inevitable Future*, Winnipeg: Arbeiter Ring Publisher.

Fazzini, K. (2019) "Israel Says It Bombed Hamas Compound That Committed Cyberattacks," *CNBC.com*, 6 May 2019. Online. Available: <www.cnbc.com/2019/05/06/israel-conflict-live-response-to-a-cyberattack-will-lead-to-a-shift.html?fbclid=IwAR0PJ3J0dlA-kZqojMq53wggpCXQYfykxLoDUdPo5udHs5sbXAVO1YEM6Pc> (accessed 6 May 2019).

Fischerkeller, M.P. and Harknett, R.J. (2017) "Deterrence Is Not a Credible Strategy for Cyberspace," *Orbis*, 61, 3: 381–93. doi:10.1016/j.orbis.2017.05.003.

Freedman, L. (1989) *The Evolution of Nuclear Strategy*, New York: St. Martin's Press. doi:10.1007/978-1-349-20165-5.

Furedi, F. (2006) *Culture of Fear Revisited: Risk-Taking and the Morality of Low Expectation*, London: Continuum.

———. (2009) "Precautionary Culture and the Rise of Possibilistic Risk Assessment," *Erasmus Law Review*, 2, 2: 197–220.

Galeotti, M. (2016) *Hybrid War Or Gibridnaya Voina? Getting Russia's Non-Linear Military Challenge Right*, London: Mayak Intelligence.

Gallagher, R. and Greenwald, G. (2014) "How the NSA Plans to Infect 'Millions' of Computers With Malware," *The Intercept*, 12 March 2014. Online. Available: <https://firstlook.org/theintercept/2014/03/12/nsa-plans-infect-millions-computers-malware/> (accessed 12 March 2014).

Geary, J. (2011) *I Is an Other: The Secret Life of Metaphor and How It Shapes the Way We See the World*, New York: HarperCollins.

Geller, E. (2019) "Trump Officials Weigh Encryption Crackdown," *Politico*, 27 June 2019. Online. Available:<www.politico.com/story/2019/06/27/trump-officials-weigh-encryption-crackdown-1385306> (accessed 27 June 2019).

Glassner, B. (1999) *The Culture of Fear: Why Americans Are Afraid of the Wrong Things*, New York, NY: Basic Books.

Goldman, E.O. and Arquilla, J. (eds) (2014) *Cyber Analogies*, Monterey, CA: Naval Postgraduate School. doi:10.21236/ada601645.

Goodin, D. (2019) "Stolen NSA Hacking Tools Were Used in the Wild 14 Months Before Shadow Brokers Leak," *ArsTechnica*, 7 May 2019. Online. Available: <https://arstechnica.com/information-technology/2019/05/stolen-nsa-hacking-tools-were-used-in-the-wild-14-months-before-shadow-brokers-leak/> (accessed 7 May 2019).

Gorman, S. and Barnes, J. (2011) "Cyber Combat: Act of War," *Wall Street Journal*, 31 May 2011. Online. Available: <https://online.wsj.com/article/SB10001424052702304563104 576355623135782718.html> (accessed 31 May 2011).

Greenberg, A. (2019) "How Not to Prevent a Cyberwar With Russia," *Wired*, 18 June 2019. Online. Available: <www.wired.com/story/russia-cyberwar-escalation-power-grid/> (accessed 18 June 2019).

Handmer, J. and James, P. (2007) "Trust Us and Be Scared: The Changing Nature of Contemporary Risk," *Global Society*, 21, 1: 119–30.

Hardt, M. and Negri, A. (2004) *Multitude: War and Democracy in the Age of Empire*, New York: The Penguin Press.

Hasian, J.M., Jr. (2005) *In the Name of Necessity: Military Tribunals and the Loss of American Civil Liberties*, Tuscaloosa: University of Alabama Press.

Hasian, J.M. and Lawson, S. (2018) "The Syrian Rebellion and the 'First Social Media' War," in Dee, J. (ed) *From Tahrir Square to Ferguson: Social Networks as Facilitators of Social Movements*, New York: Peter Lang, pp. 131–58. doi:10.3726/978-1-4539-1757-2.

Hasian, J.M., Lawson, S.T. and McFarlane, M. (2015) *The Rhetorical Invention of America's National Security State*, Lanham, MA: Lexington Books.

Healey, J. (2018a) "US Cyber Command: 'When Faced With a Bully . . . Hit Him Harder'," *The Cipher Brief*, 26 February 2018a. Online. Available: <www.thecipherbrief.com/column_article/us-cyber-command-faced-bully-hit-harder> (accessed 26 February 2018a).

———. (2018b) "Triggering the New Forever War, in Cyberspace," *The Cipher Brief*, 1 April 2018b. Online. Available: <www.thecipherbrief.com/triggering-new-forever-war-cyberspace> (accessed 1 April 2018b).

———. (2018c) "Not the Cyber Deterrence the United States Wants," *Net Politics*, 11 June 2018c. Online. Available: <www.cfr.org/blog/not-cyber-deterrence-united-states-wants> (accessed 11 June 2018c).

———. (2019) "Taking Down Russian Trolls Is My Kind of Cyber Attack," *The Cipher Brief*, 28 February 2019. Online. Available: <www.thecipherbrief.com/column/cyber-initiator/taking-down-russian-trolls-is-my-kind-of-cyber-attack> (accessed 28 February 2019).

Herath, T. and Rao, H.R. (2009) "Protection Motivation and Deterrence: A Framework for Security Policy Compliance in Organisations," *European Journal of Information Systems*, 18, 2: 106–25. doi:10.1057/ejis.2009.6.

Hibbitts, B.J. (1994) "Making Sense of Metaphors: Visuality, Aurality and the Reconfiguration of American Legal Discourse," *Cardozo Law Review*, 16: 229–356.

Hodge, N. and Ilyushina, M. (2019) "Putin Signs Law to Create an Independent Russian Internet," *CNN*, 1 May 2019. Online. Available: <www.cnn.com/2019/05/01/europe/vladimir-putin-russian-independent-internet-intl/index.html> (accessed 1 May 2019).

House Permanent Select Committee on Intelligence. (2015) *Statement for the Record: Worldwide Cyber Threats*, United States House of Representatives, 10 September 2015.

Hughes, C.R. (2002) "China and the Globalization of ICTs: Implications for International Relations," *New Media & Society*, 4, 2: 205–24. doi:10.1177/14614440222226343.

Janson, J.P. (2018) "It's Time to Take the Human Domain Seriously. USCYBERCOM Is Our Chance," *OTH Journal*, 18 May 2018. Online. Available: <https://othjournal.com/2018/05/18/its-time-to-take-the-human-domain-seriously-uscybercom-is-our-chance/> (accessed 18 May 2018).

Jarvis, J. (2011) "Revealed: Us Spy Operation That Manipulates Social Media," *The Guardian*, 17 March 2011. Online. Available: <www.theguardian.com/technology/2011/mar/17/us-spy-operation-social-networks> (accessed 17 March 2011).

Jensen, B. (2019) "What a U.S. Operation in Russia Shows About the Limits of Coercion in Cyber Space," *War on the Rocks*, 20 June 2019. Online. Available: <https://warontherocks.com/2019/06/what-a-u-s-operation-in-russia-shows-about-the-limits-of-coercion-in-cyber-space/> (accessed 20 June 2019).

Jervis, R. (1976) *Perception and Misperception in International Politics*, Princeton, NJ: Princeton University Press. doi:10.1515/9781400885114.

Johnson, D.B. (2019) "New Defense Cyber Commission Plans for Future Threats," *Defense Systems*, 13 May 2019. Online. Available: <https://defensesystems.com/articles/2019/05/15/cyber-mission-solarium.aspx> (accessed 13 May 2019).

Kaplan, F. (2016) *Dark Territory: The Secret History of Cyber War*, New York: Simon & Schuster.

Keller, E.F. (1995) *Refiguring Life: Metaphors of Twentieth-Century Biology*, New York: Columbia University Press. doi:10.7312/kell92562.

Khong, Y.F. (1992) *Analogies at War: Korea, Munich, Dien Bien Phu, and the Vietnam Decisions of 1965*, Princeton, NJ: Princeton University Press.

Kofman, M. and Rojansky, M. (2015) "A Closer Look at Russia's 'Hybrid War'," *Kennan Cable*, April, 7: 1–8.

Kok, G., et al. (2018) "Ignoring Theory and Misinterpreting Evidence: The False Belief in Fear Appeals," *Health Psychology Review*, 12, 2: 111–25. doi:10.1080/17437199.2017.1415767.

Kollars, N. and Schneider, J. (2018) "Defending Forward: The 2018 Cyber Strategy Is Here," *War on the Rocks*, 20 September 2018. Online. Available: <https://warontherocks.com/2018/09/defending-forward-the-2018-cyber-strategy-is-here/> (accessed 20 September 2018).

Konvitz, J.W. (1990) "Why Cities Don't Die: The Surprising Lessons of Precision Bombing in World War Ii and Vietnam," *American Heritage Invention & Technology Magazine*, 5, 3: 58–63.

Lakoff, A. (2006) *From Disaster to Catastrophe: The Limits of Preparedness*, Presentation. Understanding Katrina: Perspectives from the Social Sciences, the Forum of the Social Science Research Council. Online. Available: <https://understandingkatrina.ssrc.org/Lakoff/>.

Lakoff, G. and Johnson, M. (1980) *Metaphors We Live By*, Chicago: University of Chicago Press. doi:10.7208/chicago/9780226470993.001.0001.

Lamond, G. (2008) "Precedent and Analogy in Legal Reasoning," in Zalta, E.N. (ed) *The Stanford Encyclopedia of Philosophy*, Palo Alto, CA: Metaphysics Research Lab, Stanford University. doi:10.1109/jcdl.2003.1204899.

Lapointe, A. (2011) "When Good Metaphors Go Bad: The Metaphoric 'Branding' of Cyberspace," *Center for Strategic and International Studies*.

Lawson, S.T. (2011a) "Articulation, Antagonism, and Intercalation in Western Military Imaginaries," *Security Dialogue*, 42, 1: 39–56. doi:10.1177/0967010610393775.

———. (2011b) "Surfing on the Edge of Chaos: Nonlinear Science and the Emergence of a Doctrine of Preventive War in the US," *Social Studies of Science*, 41, 4: 563–84. doi:10.1177/0306312711402866.

———. (2011c) "HBGary Hearts Apple," *Forbes.com*, 22 February 2011c. Online. Available: <www.forbes.com/sites/seanlawson/2011/02/22/hbgary-hearts-apple/> (accessed 22 February 2011c).

———. (2011d) "Phantom Cyber Wars Are a Distraction," *Forbes.com*, 4 November 2011d. Online. Available: <www.forbes.com/sites/seanlawson/2011/11/04/phantom-cyber-wars-are-a-distraction/> (accessed 4 November 2011d).

———. (2012a) "Putting the 'War' in Cyberwar: Metaphor, Analogy, and Cybersecurity Discourse in the United States," *First Monday*, 17, 7. doi:10.5210/fm.v17i7.3848.

———. (2012b) "Of Cyber Doom, Dots, and Distractions," *Forbes.com*, 16 October 2012b. Online. Available: <www.forbes.com/sites/seanlawson/2012/10/16/of-cyber-doom-dots-and-distractions/> (accessed 16 October 2012b).

———. (2014) *Nonlinear Science and Warfare: Chaos, Complexity, and the U.S. Military in the Information Age*, London: Routledge. doi:10.4324/9780203766446.

———. (2015a) "Bringing the 'Cyber' Back Into U.S. Cyberwar Discourse," *Georgetown Journal of International Affairs*, October: 212–22.

———. (2015b) "With Drone Strike on ISIS Hacker U.S. Escalates Its Response to Cyber Attacks," *Forbes*, 12 September 2015b. Online. Available: <www.forbes.com/sites/seanlawson/2015/09/12/with-drone-strike-on-isis-hacker-u-s-escalates-its-response-to-cyber-attacks/#72f43066b6a8> (accessed 12 September 2015b).

———. (2015c) "Officials Seize on Paris Attacks to Push Cybersecurity Measures," *Forbes.com*, 5 December 2015c. Online. Available: <www.forbes.com/sites/seanlawson/2015/12/05/officials-seize-on-paris-attacks-to-push-cybersecurity-measures/#787fe2fabf86> (accessed 5 December 2015c).

———. (2015d) "Has ISIS Become the Top Cyber Threat?" *Forbes.com*, 18 December 2015d. Online. Available: <www.forbes.com/sites/seanlawson/2015/12/18/has-isis-become-the-top-cyber-threat/#2b4fdb03159a> (accessed 18 December 2015d).

———. (2016) "On This Date in Cyber Doom History: An Example of Getting It So Wrong for So Long," *Forbes.com*, 25 June 2016. Online. Available: <www.forbes.com/sites/seanlawson/2016/06/25/on-this-date-in-cyber-doom-history-an-example-of-getting-it-so-wrong-for-so-long/#d061e9f56b71> (accessed 25 June 2016).

Lawson, S. (2012c) "Is the united states militarizing cyberspace?," Forbes.com, 02 November 2012. Online. Available: <www.forbes.com/sites/seanlawson/2012/11/02/is-the-united-states-militarizing-cyberspace/> (accessed 02 November 2012).

Lawson, S.T. and Middleton, M.K. (2019) "Cyber Pearl Harbor: Analogy, Fear, and the Framing of Cyber Security Threats in the United States, 1991–2016," *First Monday*, 24, 3. doi:10.5210/fm.v24i3.9623.

Lawson, S.T., et al. (2016) "The Cyber-Doom Effect: The Impact of Fear Appeals in the Us Cyber Security Debate," in Pissandis, N., Roigas, H. and Veenendaal, M. (eds) *Proceedings of the 8th International Conference on Cyber Conflict (Cycon)*, IEEE, Tallinn, Estonia: pp. 65–80. doi:10.1109/CYCON.2016.7529427.

Lee, D., Larose, R. and Rifon, N. (2008) "Keeping Our Network Safe: A Model of Online Protection Behaviour," *Behaviour & Information Technology*, 27, 5: 445–54. doi:10.1080/01449290600879344.

Lewis, J.A. (2010) "The Cyber War Has Not Begun," unpublished manuscript.

———. (2017) "The Truth About a Cyber Pearl Harbor," *CNN*, 29 August 2017. Online. Available: <www.cnn.com/2017/08/29/opinions/truth-about-cyber-pearl-harbor-lewis/index.html> (accessed 29 August 2017).

Libicki, M.C. (1997) *Defending Cyberspace, and Other Metaphors*, Washington, DC: U.S. Government Printing Office.

———. (2009) *Cyberdeterrence and Cyberwar*, Santa Monica, CA: RAND.

Lipton, E., Sanger, D.E. and Shane, S. (2016) "The Perfect Weapon: How Russian Cyberpower Invaded the US," *New York Times*, 13 December 2016. Online. Available: <www.nytimes.com/2016/12/13/us/politics/russia-hack-election-dnc.html?nytmobile=0&_r=1> (accessed 13 December 2016).

Maddux, J.E. and Rogers, R.W. (1983) "Protection Motivation and Self-Efficacy: A Revised Theory of Fear Appeals and Attitude Change," *Journal of Experimental Social Psychology*, 19, 5: 469–79. doi:10.1016/0022-1031(83)90023-9.

Markoff, J. and Shanker, T. (2009) "Panel Advises Clarifying U.S. Plans on Cyberwar," *New York Times*, 30 April 2009. Online. Available: <www.nytimes.com/2009/04/30/science/30cyber.html?_r=1> (accessed 30 April 2009).

Maxey, L. (2017) "U.S. Searches for Cyber Doctrine with Russians 'Ten Years Ahead'," *The Cipher Brief*, 22 September 2017. Online. Available: <www.thecipherbrief.com/article/exclusive/tech/u-s-searches-cyber-doctrine-russians-ten-years-ahead> (accessed 22 September 2017).

McConnell, M. (2010) "Mike McConnell on How to Win the Cyber-war We're Losing," *The Washington Post*, 28 February 2010, B01.

McCullagh, D. (2010) "Senators Propose Granting President Emergency Internet Power," *CNET News*, 10 June 2010. Online. Available: <https://news.cnet.com/8301-13578_3-20007418-38.html> (accessed 10 June 2010).

McPherson, P. (2019) "Trade Deal or Not, US Must Counter China Moves to Beat Us at Science and Technology," *USA Today*, 20 May 2019. Online. Available: <www.usatoday.com/story/opinion/2019/05/20/china-us-scientific-research-technology-innovation-column/3685672002/> (accessed 20 May 2019).

Mongoven, A. (2006) "The War on Disease and the War on Terror: A Dangerous Metaphorical Nexus?" *Cambridge Quarterly of Healthcare Ethics*, 15, 4: 403–16. doi:10.1017/s0963180106060518.

Mullen, A.D.M.M. (2009) "Strategic Communication: Getting Back to Basics," *Joint Forces Quarterly*, 55, 4: 2–4.

Nafeez, A. (2016) "Your Government Wants to Militarize Social Media to Influence Your Beliefs," *Motherboard*, 14 November 2016. Online. Available: <www.vice.com/en_us/article/9a384v/your-government-wants-to-militarize-social-media-to-influence-your-beliefs> (accessed 14 November 2016).

Nakashima, E. (2010) "Dismantling of Saudi-CIA Web Site Illustrates Need for Clearer Cyberwar Policies," *The Washington Post*, 19 March 2010. Online. Available: <www.washingtonpost.com/wp-dyn/content/article/2010/03/18/AR2010031805464_pf.html> (accessed 19 March 2010).

Nerhot, P. (1991) *Legal Knowledge and Analogy: Fragments of Legal Epistemology, Hermeneutics, and Linguistics*, Boston: Kluwer Academic Publishers. doi:10.1007/978-94-011-3260-2.

Noon, D.H. (2004) "Operation Enduring Analogy: World War II, the War on Terror, and the Uses of Historical Memory," *Rhetoric & Public Affairs*, 7, 3: 339–64. doi:10.1353/rap.2005.0015.

Norman, P. (2019) "U.S. Unleashes Military to Fight Fake News, Disinformation," *Bloomberg*, 31 August 2019. Online. Available: <www.bloomberg.com/news/articles/2019-08-31/u-s-unleashes-military-to-fight-fake-news-disinformation> (accessed 31 August 2019).

Ortony, A. (ed) (1979) *Metaphor and Thought*, Cambridge: Cambridge University Press. doi:10.1017/CBO9781139173865.

Owens, W.A., Dam, K.W. and Lin, H.S. (2009) *Technology, Policy, Law, and Ethics Regarding U.S. Acquisition and Use of Cyberattack Capabilities*, Washington, DC: National Academies Press. doi:10.17226/12651.

Pagliery, J. and Perez, E. (2015) "Super-sneaky Malware Found in Companies Worldwide," *CNN.com*, 17 February 2015. Online. Available: <https://money.cnn.com/2015/02/17/technology/security/malware-nsa/> (accessed 17 February 2015).

Paparone, C.R. (2008) "Metaphors We Are Led By," *Military Review*, November–December: 55–64.

Perloth, N., Sanger, D.E. and Shane, S. (2019) "How Chinese Spies Got the NSA's Hacking Tools, and Used Them for Attacks," *The New York Times*, 6 May 2019. Online. Available: <www.nytimes.com/2019/05/06/us/politics/china-hacking-cyber.html> (accessed 6 May 2019).

Perloth, N. and Shane, S. (2019) "In Baltimore and Beyond, a Stolen NSA Tool Wreaks Havoc," *The New York Times*, 25 May 2019. Online. Available: <www.nytimes.com/2019/05/25/us/nsa-hacking-tool-baltimore.html> (accessed 25 May 2019).

Peters, G.-J.Y., Ruiter, R.A.C. and Kok, G. (2013) "Threatening Communication: A Critical Re-Analysis and a Revised Meta-Analytic Test of Fear Appeal Theory," *Health Psychology Review*, 7, sup1: S8–S31. doi:10.1080/17437199.2012.703527.

———. (2014) "Threatening Communication: A Qualitative Study of Fear Appeal Effectiveness Beliefs Among Intervention Developers, Policymakers, Politicians, Scientists, and Advertising Professionals," *International Journal of Psychology*, 49, 2: 71–9. doi:10.1002/ijop.12000.

Pfau, M. (2007) "Who's Afraid of Fear Appeals? Contingency, Courage, and Deliberation in Rhetorical Theory and Practice," *Philosophy and Rhetoric*, 40, 2: 216–37. doi:10.1353/par.2007.0024.

Pfleeger, S.L. and Caputo, D.D. (2012) "Leveraging Behavioral Science to Mitigate Cyber Security Risk," *Computers & Security*, 31, 4: 597–611. doi:10.1016/j.cose.2011.12.010.

Pomerleau, M. (2018a) "Where Do Information Operations Fit in the DOD Cyber Enterprise?" *Fifth Domain*, 26 July 2018a. Online. Available: <www.fifthdomain.com/c2-comms/2018/07/26/where-do-information-operations-fit-in-the-dod-cyber-enterprise/> (accessed 26 July 2018a).

———. (2018b) "Is the Defense Department's Entire Vision of Cybersecurity Wrong?," *Fifth Domain*, 16 November 2018b. Online. Available: <www.fifthdomain.com/dod/2018/11/14/is-the-defense-departments-entire-vision-of-cybersecurity-wrong/> (accessed 16 November 2018b).

Poulsen, K. (2003) "Fed: Cyberterror fears Missed Real Threat: Effort Was 'Misdirected'," *The Register*, 2003. Online. Available: <www.theregister.co.uk/2003/08/01/fed_cyberterror_fears_missed_real/> (accessed 2003).

Pratt, M.K. (2019) "How to Market Security: 8 Tips for Recruiting Users to Your Cause," *CSO Online*, 19 August 2019. Online. Available: <www.csoonline.com/article/3430803/how-to-market-security-8-tips-for-recruiting-users-to-your-cause.html> (accessed 19 August 2019).

Quarantelli, E.L. (2008) "Conventional Beliefs and Counterintuitive Realities," *Social Research: An International Quarterly*, 75, 3: 873–904.

Raduege (Ret), L.G.H.R. (2011) "Deterring Attackers in Cyberspace," *The Hill*, 23 September 2011. Online. Available: <https://thehill.com/opinion/op-ed/183429-deterring-attackers-in-cyberspace> (accessed 23 September 2011).

Rid, T. and Buchanan, B. (2018) "Hacking Democracy," *SAIS Review of International Affairs*, 38, 1: 3–16. doi:10.1353/sais.2018.0001.

Rid, T. and McBurney, P. (2012) "Cyber-Weapons," *The RUSI Journal*, 157, 1: 6–13. doi:10.1080/03071847.2012.664354.

Rogers, A.D.M.M. and Weinstein, D. (2019) "Protecting Our Critical Infrastructure in the Digital Age," *The Hill*, 9 June 2019. Online. Available: <https://thehill.com/opinion/cybersecurity/447596-protecting-our-critical-infrastructure-in-the-digital-age> (accessed 9 June 2019).

Rogers, R.W. (1975) "A Protection Motivation Theory of Fear Appeals and Attitude Change," *The Journal of Psychology*, 91, 1: 93–114. doi:10.1080/00223980.1975.9915803.

Sanger, D.E. and Perloth, N. (2019) "U.S. Escalates Online Attacks on Russia's Power Grid," *The New York Times*, 15 June 2019. Online. Available: <www.nytimes.com/2019/06/15/us/politics/trump-cyber-russia-grid.html> (accessed 15 June 2019).

Sanger, D.E. and Schmitt, E. (2017) "U.S. Cyberweapons, Used Against Iran and North Korea, Are a Disappointment Against ISIS," *The New York Times*, 12 June 2017. Online. Available: <www.nytimes.com/2017/06/12/world/middleeast/isis-cyber.html> (accessed 12 June 2017).

Saperstein, A.M. (1997) "Complexity, Chaos, and National Security Policy: Metaphors or Tools?" in Alberts, D.S. and Czerwinski, T.J. (eds) *Complexity, Global Politics, and National Security*, Washington, DC: National Defense University, pp. 44–61.

Schneier, B. (2014) "Quantum Technology Sold by Cyberweapons Arms Manufacturers," *Schneier on Security*, 14 August 2014. Online. Available: <www.schneier.com/blog/archives/2014/08/quantum_technol.html> (accessed 14 August 2014).

Schön, D.A. (1993) "Generative Metaphor: A Perspective on Problem-Setting in Social Policy," in Ortony, A. (ed) *Metaphor and Thought*, Cambridge: Cambridge University Press, pp. 137–63. doi:10.1017/cbo9781139173865.

Schulte, S.R. (2013) *Cached: Decoding the Internet in Global Popular Culture*, New York: New York University Press. doi:10.18574/nyu/9780814708668.003.

Seablom, D. and Helms, N. (2018) "Fighting the Wrong Fight," *OTH Journal*, 1 August 2018. Online. Available: <https://othjournal.com/2018/08/01/fighting-the-wrong-fight/> (accessed 1 August 2018).

Sefat, E., et al. (2017) "Deterministic Thinking and Mental Health: A Review Article," *British Journal of Education, Society & Behavioural Science*, 19, 2: 1–10. doi:10.9734/bjesbs/2017/28972.

Sherman, J. (2019) "Russia and Iran Plan to Fundamentally Isolate the Internet," *Wired*, 6 June 2019. Online. Available: <www.wired.com/story/russia-and-iran-plan-to-fundamentally-isolate-the-internet/> (accessed 6 June 2019).

Shields, P. (2005) "When the 'Information Revolution' and the US Security State Collide: Money Laundering and the Proliferation of Surveillance," *New Media & Society*, 7, 4: 483–512. doi:10.1177/1461444805054110.

Singer, P.W. (2014) *Cybersecurity and Cyberwar: What Everyone Needs to Know*, London: Oxford University Press.

Siponen, M., Adam, M. and Pahnila, S. (2014) "Employees' Adherence to Information Security Policies: An Exploratory Field Study," *Information & Management*, 51, 2: 217–24. doi:10.1016/j.im.2013.08.006.

Smeets, M. (2018) "US Cyber Command: An Assiduous Actor, Not a Warmongering Bully," *The Cipher Brief*, 4 March 2018. Online. Available: <www.thecipherbrief.com/article/exclusive/tech/us-cyber-command-assiduous-actor-not-warmongering-bully> (accessed 4 March 2018).

Stone, A. (2019) "How Leon Panetta's 'Cyber Pearl Harbor' Warning Shaped Cyber Command," *Fifth Domain*, 30 July 2019. Online. Available: <www.fifthdomain.com/opinion/2019/07/30/how-leon-panettas-cyber-pearl-harbor-warning-shaped-cyber-command/> (accessed 30 July 2019).

Sunstein, C.R. (2002) "Probability Neglect: Emotions, Worst Cases, and Law," *The Yale Law Journal*, 112, 1: 61–107. doi:10.2307/1562234.

———. (2005) *Laws of Fear: Beyond the Precautionary Principle*, Cambridge: Cambridge University Press. doi:10.1017/cbo9780511790850.

———. (2007) *Worst-Case Scenarios*, Cambridge, MA: Harvard University Press. doi:10.4159/9780674033535.

Valeriano, B. and Jensen, B. (2019) "The Myth of the Cyber Offense: The Case for Restraint," *CATO Policy Analysis*, 862: 1–16.

Valeriano, B., Roff, H. and Lawson, S. (2016) "Dropping the Cyber Bomb? Spectacular Claims and Unremarkable Effects," *Net Politics*, 24 May 2016. Online. Available: <www.cfr.org/blog/dropping-cyber-bomb-spectacular-claims-and-unremarkable-effects> (accessed 24 May 2016).

van der Ploeg, I. (2003) "Biometrics and Privacy a Note on the Politics of Theorizing Technology," *Information, Communication & Society*, 6, 1: 85–104. doi:10.1080/136911803 2000068741.

von Ghyczy, T. (2003) "The Fruitful Flaws of Strategy Metaphors," *Harvard Business Review*, September.

Wall, D.S. (2008) "Cybercrime and the Culture of Fear: Social Science Fiction(s) and the Production of Knowledge About Cybercrime," *Information, Communication & Society*, 11, 6: 861–84. doi:10.1080/13691180802007788.

Walton, D.N. (2000) *Scare Tactics: Arguments That Appeal to Fear and Threats*, Boston: Kluwer Academic Publishers. doi:10.1007/978-94-017-2940-6.

Walworth, A. (2018) "The Perils of Confusing Nuclear and Cyber Strategy," *Real Clear Politics*, 12 July 2018. Online. Available: <www.realclearpolitics.com/articles/2018/07/12/the_perils_of_confusing_nuclear_and_cyber_strategy_137500.html> (accessed 12 July 2018).

Webster, S.C. (2011) "Revealed: Air Force Ordered Software to Manage Army of Fake Virtual People," *Raw Story*, 18 February 2011. Online. Available: <www.rawstory.com/2011/02/revealed-air-force-ordered-software-to-manage-army-of-fake-virtual-people/> (accessed 18 February 2011).

Weinreb, L.L. (2005) *Legal Reason: The Use of Analogy in Legal Argument*, Cambridge: Cambridge University Press. doi:10.1017/cbo9780511810053.

Witte, K. (1994) "Fear Control and Danger Control: A Test of the Extended Parallel Process Model (EPPM)," *Communication Monographs*, 61, 2: 113–34. doi:10.1080/03637759409376328.

Wyatt, S. (2004) "Danger! Metaphors at Work in Economics, Geophysiology, and the Internet," *Science, Technology & Human Values*, 29, 2: 242–61. doi:10.1177/0162243903261947.

Zetter, K. (2009) "Lawmaker Wants 'Show of Force' Against North Korea for Website Attacks," *Wired Threat Level*, 10 July 2009. Online. Available: <www.wired.com/threatlevel/2009/07/show-of-force/> (accessed 10 July 2009).

———. (2014) *Countdown to Zero Day: Stuxnet and the Launch of the World's First Digital Weapon*, New York: Crown Publishers.

6 Cold War 2.0 and the emergence of cyber-enabled political warfare

Introduction

As noted throughout the preceding chapters, critics have argued for years that U.S. obsession with cyber-doom was a potentially dangerous distraction. Even the most severe examples of cyber conflict to date have not exhibited the kinds of damage contemplated in cyber-doom rhetoric. This fact was becoming apparent on the eve of the Russian cyber operations against the 2016 U.S. presidential election. As noted in Chapter 5, U.S. cyber operations against ISIS helped to spark a shift in U.S. military thinking about how cyber capabilities could be used most effectively. As we saw in Chapter 4, even prior to 2016, various observers were pointing to Russian cyber operations in the Ukraine conflict as evidence that cyber war was not turning out the way we had imagined. Russian cyber interference in the 2016 U.S. presidential election, combined with revelations that other nation states are already developing and deploying similar capabilities (Landay and Hosenball, 2018), has been a watershed moment for shifting U.S. thinking about cyber conflict. But where do we go from here and how can we do better? To begin addressing that question, this chapter critically examines the post-2016 debate in the United States over how to think and speak about the kinds of cyber-enabled information operations used by Russia in 2016.

As discussed in Chapter 1, the formation of security imaginaries involves the use of various rhetorical tactics to articulate particular situations and their component parts as threatening (Hasian et al., 2015). In Chapter 2, we saw that this articulation includes the use of analogies and metaphors, both historical and otherwise (Makus, 1990), as well as a number of other rhetorical tactics. However, certain events perceived to be sudden or unprecedented – "focusing" or "signal" events – can serve as antagonisms that unsettle dominant articulations (Lawson, 2011). A shift in language use can be an indicator that dominant articulations have become unstable. This can include the sudden emergence of new terms, re-emergence of older terms, or confusion and debate about which terms and concepts to use (Doyle, 1997: 10).

This is precisely what we see in the wake of 2016. More than three years after the attacks were first revealed, there remains a lack of clarity about just what this incident signals about the nature and future of cyber conflict. This is evidenced by the conceptual and terminological confusion that can be found in the ongoing debate. Some commentators refer to the Russian operation as an example of "hybrid warfare," others as "cyber warfare," and still others as "information warfare" or "political warfare." Some use a combination of these, often with little explanation of what they mean or how they relate to one another. Despite the fact that most of these terms are decades old, there is a tendency to depict what Russia carried out in 2016 – and before that in Ukraine and elsewhere – as the emergence of a dangerous new form of warfare. In short, in wrestling with how to describe the 2016 Russia operation, scholars, commentators, and officials are often unsure just what the terms and concepts being used mean, how they relate to one another, whether Russian operations fit the definition of one or more of them, and whether any of this is new or even war.

This chapter describes and critically evaluates these efforts to play rhetorical and conceptual catch-up. The next section demonstrates the usage of a number of different terms and concepts by policymakers and experts. The remaining sections compare the major features of what each of these terms and concepts – cyber, information, hybrid, and political warfare – have meant historically in the United States. Together, they argue that the resurgence of concepts like information and political warfare is the result of the dominance of a narrow understanding of cyber warfare in the years leading up to the 2016 Russian election interference, which left observers with inadequate language and concepts to describe what took place in 2016. This is especially the case if we conceive of Russia's 2016 U.S. election interference as part of a broader conflict with the West that was ongoing for years prior to the 2016 election. Thus, this chapter argues that the concept of political warfare as articulated in the early years of the Cold War provides the best description of Russian activities. Ironically, then, it affirms the tendency in the U.S. cybersecurity debate to look to the Cold War for analogies to guide our thinking. However, instead of nuclear weapons and deterrence, the emergence of technologies of mass communication and strategies of political warfare may provide more apt starting points for thinking about the future of cyberspace operations in international conflict.

Warfare by any other name

Writing in *The Washington Post* in July 2018, Professor Brian Klaas implored readers to "stop calling it 'meddling.' It's actually information warfare." This shift in language use is necessary, he claimed, to "treat the threat with the seriousness it deserves." Russia's "attacks," he said, are "a part of *gibridnaya voina* – Russian for 'hybrid warfare.' The best term for what we're talking about would be 'information warfare'" (Klaas, 2018). The first implication is that without a militarized framing of the Russian actions as "attacks" and "warfare," we cannot take the threat seriously enough. The second is that "hybrid warfare" is a Russian term and that it is equivalent to what we would call "information warfare." As we will see later in this chapter, "hybrid warfare" is neither a Russian concept, nor equivalent to information warfare.

We can find many other examples from commentators of various kinds trying to make sense of Russian actions using a variety or terms and concepts. Maness and Jaitner, for example, claim that Russian "cyber operations are a subset of overall information warfare" (Maness and Jaitner, 2018). Valeriano and Jensen call the Russian operation both an "information warfare campaign" and "a modern twist to political warfare." Like others, whether information or political warfare, in their view, "Cyber was one component of a larger coercive campaign – more than espionage, but less than war" (Valeriano and Jensen, 2017). Former FBI Special Agent Clint Watts prefers a combination of "psychological warfare" and "information warfare" to describe Russian operations spanning from the 2007 DDOS attacks on Estonia discussed in Chapter 4 to interference in the 2016 U.S. presidential election (Watts, 2018).

Others go further in emphasizing the supposedly war-like nature of Russian operations in cyberspace. Historian Timothy Snyder has lamented the use of the term "hybrid war" to describe Russian actions in Ukraine, saying the term implies that they were somehow less than war when, in reality, they were "war plus" because they involved "regular war," "a partisan campaign," and "the broadest cyber offensive in history" (Snyder, 2018: 223). He goes on to call Russian actions in Ukraine and in the 2016 U.S. election both "information war" and "cyberwar" on numerous occasions. For example, he argues, "The aim of Russian cyberwar was to bring Trump to the Oval Office" (Snyder, 2018: 229). This "cyberwar" included the use of Russian "cyberweapons" carrying various "payloads," including the "fiction of 'Donald Trump, successful businessman'" and even Trump himself, according to Snyder (Snyder, 2018: 283, 252).

But it is not just pundits and academics who have deployed a variety of terms and concepts to make sense of Russian actions in 2016 and beyond. Public officials have also deployed a range of sometimes contradictory terms to label what the Russians have been doing. In a speech to the Atlantic Council one week before his stint as National Security Advisor ended, Lt. Gen. H.R. McMaster called the broad sweep of Russian actions against its neighbors and the West "hybrid warfare." This, he defined as

> a pernicious form of aggression that combines political, economic, informational, and cyber assaults against sovereign nations. Russia employs sophisticated strategies deliberately designed to achieve objectives while falling below the target state's threshold for a military response. Tactics include infiltrating social media, spreading propaganda, weaponizing information, and using other forms of subversion and espionage. [. . .] Russia brazenly and implausibly denies its actions.
>
> (McMaster, 2018)

This definition would seem to contradict Professor Klaas' earlier, implying that information "assaults" are just one part of a wider "hybrid warfare." As for the novelty of Russia's approach, McMaster said that it is a "new form of Soviet-era active measures and *maskirovka*" (McMaster, 2018) – i.e. both new and old.

Other public officials have not been as sure about what to call Russian actions, or whether they are really war or even new. For example, there was noticeable disagreement and confusion in the use of terms during the March 2017 Congressional hearings in which then-FBI Director James Comey and NSA Director ADM Michael Rogers testified. Asked to describe what Congressman Frank LoBiondo (R-NJ) called "active measures" during the election, ADM Rogers said that "the difference *this time* was the cyber dimension," which was just one part of a larger effort (emphasis added). He noted that the use of cyber tactics in support of active measures was new, but not the use of active measures per se. Later, Congresswoman Elise Stefanik (R-NY) agreed with ADM Rogers, noting that "cyber hacking" was "just one tactic" used in a "broader influence or information warfare campaign." Thus, they seemed to be in agreement that "cyber" is part of something bigger, but just what to call that something – active measures or information warfare – was unclear.

Two other members of Congress floated the idea that this something bigger should be called "hybrid warfare," to which Director Comey and ADM Rogers both seemed ambivalent. Asked by Representative André Carson (D-IN) to describe how Russian operations in Ukraine relate to "hybrid warfare," ADM Rogers said that it is about "influence [campaigns] and attempts to distance Russian actions from any potential blowback to the Russian state." Later, Congresswoman Jackie Speier (D-CA) echoed this definition, saying that hybrid warfare "blends conventional warfare, irregular warfare and cyber warfare. The aggressor intends to avoid attribution or retribution." But when she asked Comey and Rogers if Russia had "engaged in hybrid warfare" against the United States in 2016, they both demurred. Comey replied, "I don't think I would use the term warfare. [. . .] They engaged in a multifaceted campaign of active measures," implying that the latter is something different than the former. Rogers said, simply, "I'd agree with the director." Unsatisfied with these answers, the Congresswoman commented, "I actually think that their engagement was an act of war, an act of hybrid warfare," again, the implication being that "active measures" are something different and less concerning than "hybrid warfare." Finally, Congresswoman Stefanik followed up with the two witnesses, noting once again that "the use of cyber tools [was] part of their broader, whether you call it hybrid warfare or information warfare campaigns."

Whether cyber, information, or hybrid, plenty of observers have been willing to join Professor Snyder in suggesting that the Russian actions in 2016 were an "act of war," itself a commonly-used, official-sounding term with an ambiguous relationship to more precise domestic and international legal terms like "use of force" and "armed attack." Senator Jeanne Shaheen (D-NH) commented, "We should think about whether it [Russian election interference] is an act of war or not," which a recent defense authorization bill tasked the Trump administration with figuring out (Chalfant, 2017). Former Senator John McCain (R-AZ) needed no more convincing, however, saying of the Russians' 2016 election interference, "When you attack a country, it's an act of war" (Reuters, 2016).

Of course, as with all things "cyber," the obligatory Pearl Harbor or 9/11 analogies are always waiting, as when Lt. Gen. Mark Hertling (ret) and Molly McKew told readers of *Politico* that the aim of the 2016 Russian election interference "was every bit as much to devastate the American homeland as Pearl Harbor or 9/11" and that it was largely successful (Hertling and McKew, 2018). This is, of course, another example of the kind of exaggeration that comes from applying the Pearl Harbor or 9/11 analogy to cyber operations that have already occurred, which we encountered in Chapter 2. Though certainly very serious, the impacts of the 2016 Russian cyber operations against the U.S. election do not approximate the death and destruction seen during Pearl Harbor or 9/11, nor the impacts most often contemplated in cyber-doom scenarios that draw on these historical events for their inspiration (Lawson and Middleton, 2019).

The issues that arise from these examples include how to label the Russian actions overall, but also how to label the individual components and describe how they relate to one another. We also see questions of what, if anything, is new about what the Russians have been doing and whether it should be understood as "war." Thus, how we talk about the Russian operations matters. As Russia expert Mark Galeotti wrote, "[W]ords have weight; they frame our understanding of that [Russian] campaign, of how it works, and what it does. Without being aware of it, clinging to [an] inaccurate moniker also limits and misdirects us in our attempt to grasp and thus combat it" (Galeotti, 2018).

In addition to understanding the Russians on their own terms – what the Army's Foreign Military Studies Office calls studying "how they think they think" – we must do a better job of translating this understanding into our own terms, that is, taking what the Russians have done and finding a familiar frame of reference from our own experience. That is the driving motivation of the remainder of this chapter, which examines the history of the dominant American terms and analogies used to describe the Russian operations. There is a long history of debate around these issues in the United States. If we dust off some of that material, we can improve our conceptual clarity, as well as find nuggets of wisdom to help guide our understanding and responses to the current challenges.

Cyber fires, pew pew pew!

In the years leading up to Russian interference in the 2016 U.S. presidential election, there was increasing public debate in the United States about cyber warfare. As a result, and given the centrality of computer hacking techniques to the Russian operation, it is unsurprising that some have framed the Russian action as cyber warfare. However, if we measure the 2016 Russian operation against historical U.S. understandings of cyber warfare, we can see that cyberspace operations were but one part of a larger Russian operation. While authoritative definitions of cyber warfare have shifted over time, becoming broader in some respects but narrower in others, nonetheless, the U.S. definition of "cyber" would only accurately describe a portion of the 2016 Russian operation.

RAND researchers John Arquilla and David Ronfeldt are largely credited as being first to popularize the term "cyber war" in their 1993 working paper and then 1997 essay, both titled, "Cyberwar is Coming!" In their original definition, cyber war was about traditional, kinetic military conflict. Rooted in the context of 1980s military reform and resultant post-Desert Storm debates about a supposed revolution in military affairs discussed in Chapter 3, cyber war at this time was about the growing centrality of information, knowledge, and associated technologies to military operations, including command and control, intelligence, and precision munitions (Arquilla and Ronfeldt, 1997: 30). Arquilla and Ronfeldt reserved a different term for the kind of conflict we have seen Russia waging in recent years. That term was "netwar" and it encompassed a much wider variety of activities, including public diplomacy, psychological operations, deception, propaganda, political and cultural subversion, economic warfare, and, yes, computer network intrusions. They were clear that netwar was "not real war" and that it was also distinct from cyber war. Netwar, they said, would only become cyber war if, for example, it was used to target military command and control systems (Arquilla and Ronfeldt, 1997: 28, 30).

Thus, on the questions of war or not, new or old, Arquilla and Ronfeldt were pretty clear. Cyber war is actual war; netwar is not. Conceptually, they said that cyber war was not entirely new because it resembled Thomas Rona's 1976 "concept of an 'information war' that is 'intertwined with, and superimposed on, other military operations'" (Quoting Rona, 1976: 2, Arquilla and Ronfeldt, 1997: 31). In practice, cyber war was also not entirely new, they said, precisely because it is war first, cyber second. Information had always been important in warfare, they said, but cyber war represented a shift in emphasis where information was now at the center of all warfighting activities. Cyber war in this definition was not something separate from traditional, kinetic warfare, or waged in an entirely different domain. It was the suffusion of ICTs, of cyberspace, into all aspects of traditional warfighting. It was not, as is common in today's discourse, primarily about computers attacking computers or the hacking of critical infrastructure. Similarly, they noted that while the elements of netwar were not new, their convergence in the Information Age represented "a new entry on the spectrum of conflict" (Arquilla and Ronfeldt, 1997: 28). The newness of netwar as described here is debatable, however, as we will see in the section on political warfare that follows.

This early definition of cyber war has shifted over the years among academics and defense intellectuals, but also in official U.S. policy related to cyberspace operations. Among the first group, Clarke and Knake's (2010) book, which along with Arquilla and Ronfeldt's article is one of the two most cited works on cyber war,[1] sees cyber war as including at least portions of what Arquilla and Ronfeldt called netwar, but also having the possibility to exist outside the context of traditional warfare. Clarke and Knake point to the second U.S. invasion of Iraq in 2003 and the Israeli attack on Syrian nuclear facilities in 2007 as representing "two uses of cyber war." The first, they said, was using cyber war to take out enemy defenses in support of a traditional, kinetic attack. The second "use of cyber war is to send propaganda out to demoralize the enemy." In the 2007 DDOS attacks on Estonia that were widely attributed to Russia, however, they saw the emergence of something new, what they called "stand-alone cyber war" that "does not . . . have to be accompanied by bombing raids or tank battles" (Clarke and Knake, 2010: 20–1).

Analysis of DOD and USCYBERCOM cyberspace strategy documents reveals many of the same patterns, but with some notable shifts beginning to emerge since the 2016 Russia operation. For example, they all note that cyber conflict takes place not just during war or time of crisis, but during the full spectrum of conflict, including "from Phase 0 (peacetime daily ongoing) operations, through Phase 3 (conflict and crisis), to Phase 5 (recovery) operations" (Department of Defense, 2015a). They commonly describe cyber conflict using variations on the terms "continual," "persistent," and "daily." Such conflict takes place, the 2011 DOD cyber strategy tells us, in a "man-made domain" made up of "networked systems, devices, and platforms" that combine to form "a network of networks that includes thousands of ISPs [Internet Service Providers] across the globe" (Department of Defense, 2011: 1, 9). Later documents note the "dynamic" (Department of Defense, 2015b) and "fluid" (Department of Defense, 2018a) nature of this "domain," with the 2018 USCYBERCOM Vision even noting, "The cyberspace domain that existed at the creation of US Cyber Command (USCYBERCOM) has changed" (Department of Defense, 2018a: 2). But few of these documents describe the cyberspace domain in any more detail than the 2011 DOD strategy, or say much about how the domain has supposedly changed. Likewise, the official definition of "cyberspace" in the September 2018 edition of Joint Publication (JP) 1–02, the *DOD Dictionary of Military and Associated Terms*, focuses on "networks of information technology infrastructures and resident data."

With cyberspace defined as infrastructure and data, it is unsurprising to see repeated across DOD and USCYBERCOM strategy documents some variation on the statement that we must protect, defend, or secure networks and systems, data and information. The networks, systems, data, and information to be protected are both government and, in some cases, private. Protecting these systems and data is described as a means of protecting not just our "economic and national security" (Department of Defense, 2011), but also our values, including "freedom, liberty, prosperity, intellectual property, and personal information," as well as "the free flow of information that fosters growth and intellectual dynamism" (Department of Defense, 2015a: 2).

Also of note is what is missing from these documents. At least since the creation of USCBERCOM in 2008, U.S. cyberspace operations have not been primarily about influence campaigns and shaping public opinion, or about defending the American civilian population from such operations by others (Pomerleau, 2018). Instead, as JP 1–02 tells us, "cyberspace attack" is "considered a form of fires" that is about creating denial, degradation, or destruction in cyber or physical space. Defense is about preventing that from happening to us. Cyberspace attack as "fires" can sometimes be taken quite literally, as when Deputy Secretary of Defense Bob Work told reporters that the U.S. was "dropping cyber bombs" on ISIS (Clark, 2016), a claim that was picked up and spread not just by the news media, but also other officials like Secretary of Defense Ash Carter (Valeriano et al., 2016).

However, as discussed in the last chapter, this narrow focus on attacking and defending networks and systems as "fires" is changing and is already being reflected in official military cyberspace policy. The 2018 *Department of Defense Cyber Strategy* mentions, for the first time, adversary use of cyberspace operations to "challenge our democratic processes," calling out Russia in particular for its "cyber-enabled influence operations" (Department of Defense, 2018b: 2). The proceedings from the February 2018 USCYBERCOM Cyberspace Strategy Symposium note that the United States has historically "sub-divided" cyber and information operations while adversaries have not. It therefore encouraged further exploration of "the relationship between cyber . . . and influence operations . . . and how to integrate these capabilities," even "How cyber is a subset of information operations" (Department of Defense, 2018c: 2). This view is reflected in the April 2018 USCYBERCOM vision document. Of the five "imperatives" listed in the document, one includes: "Enhance information warfare options for Joint Force commanders. Integrate cyberspace operations with information operations" (Department of Defense, 2018a: 9).

Clearly, the Russian operation in 2016 made use of cyberspace operations. But, this was not the entire story. Cyber was used to enable something bigger, which included influence and information operations. In the next section, we turn our attention to what "information warfare" and "information operations" have meant in official U.S. policy over the last three decades. We will see that though these concepts capture more of what the Russians have been doing, they are still likely too narrow to capture the full range of Russian operations against the U.S. and the West more generally.

Bits-and-bytes, hearts-and-minds

As we saw earlier, several observers have argued that the 2016 Russian operation should be called "information warfare" rather than mere "meddling" or "interference." As noted in the last section, even U.S. Cyber Command seems to be coming around to the position that cyber is something narrower than information operations. Calling the 2016 and ongoing Russian operations "information warfare" is all the easier when we learn from two different Department of Justice indictments that Russians themselves are using this term to describe their actions against the United States (Samuelsohn, 2018; Goldman, 2018). Nonetheless, that does not mean that the Russian understanding of that term is the same as our own. In this section, we explore the evolution of U.S. definitions of information warfare/operations (IW/O). Though IW/O is correctly understood today as something broader than cyberspace operations, this was not always the case. Like cyber warfare, authoritative IW/O definitions have narrowed and broadened at various points over the last three decades.

As Arquilla and Ronfeldt noted earlier, the term "information war" first appeared in 1976, in a report written by Boeing engineer Thomas Rona for DOD's famed Office of Net Assessment. Widespread debate about information warfare would not take root in the U.S. defense community until the 1990s, however. This first appearance occurred at a time when the United States sought to offset Soviet numerical superiority in conventional weapons and parity in nuclear forces with qualitative superiority through the incorporation of emerging information, communication, and computing technologies into weapons systems of all kinds (Tomes, 2007). This initial usage of "information warfare" was in line with this larger vision and, as such, focused primarily on information-enabled weapon systems (Rona, 1976). It would presage one of two competing definitions of information warfare that have coexisted, at times uneasily, in DOD policy.

Brunner and Dunn Cavelty have observed that one definition saw information warfare as primarily about attacking and defending information systems and the information stored and transmitted by them as physical infrastructures and commodities. The second took a broader view of information warfare, seeing it not merely as about attacking information systems, but about shaping perceptions, beliefs, and ultimately decisions and actions of an adversary more broadly. We might call these the bits-and-bytes versus hearts-and-minds framings of information warfare (Brunner and Dunn Cavelty, 2009). Though the two have coexisted in official DOD policy since at least 1996, debates about which should be dominant emerge periodically in response to new technological developments. Indeed, the advent of social media sparked such a debate between adherents of the bits-and-bytes and hearts-and-minds framings as far back as 2007 (Lawson, 2013).

After its first appearance in 1976, information warfare returned in the 1990s. Since then, the pendulum has swung between these two definitions, with the scope of what information warfare is said to entail becoming broader or narrower depending on the speaker and situation. On one hand, some computer security advocates like Winn Schwartau deployed the bits-and-bytes framing to warn of attacks on, or enabled by, information and information systems resulting in a "digital Pearl Harbor" (Schwartau, 1994, 1991). On the other, defense intellectuals and academics like George Stein, Martin Libicki, and Dorothy Denning presented a broader view of information warfare more in line with the hearts-and-minds framing (Stein, 1995; Libicki, 1995; Denning, 1999).

Official DOD policy on information operations has acknowledged the existence of these two, sometimes competing understandings. The 2001 version of DOD Directive 3600.1 (DODD 3600.1 2001) states this explicitly:

1   One set of IO activities employed by the DoD Components focuses on the perceptions and attitudes of decision-makers or groups.
2   A second set of IO activities also employed by the DoD Components focuses on attacking or defending the electromagnetic spectrum, information systems, and information which supports decision makers, command and control and automated responses.

(Department of Defense, 2001: 2)

Nonetheless, observance of changes in DODD 3600.1 over the years reveals that bits-and-bytes has maintained a presence in official discourse while hearts-and-minds is a more recent addition to the overall definition of information operations. For example, the original 1992 issuance of the policy made no mention of perceptions and attitudes. Instead, it focused entirely on attacking and defending information and information systems narrowly defined as data and infrastructure, respectively (Department of Defense, 1992: enclosure 1–2). That is, it sounds more like recent definitions of cyber warfare than the descriptions of Russian information warfare that we discussed earlier.

This all began to change, however, with the issuance of the 1996 version of DODD 3600.1. Though the 1996 policy still focused on attacking and defending "information and information systems," its definitions of each of those terms were broader than those in the 1992 policy. Information was no longer just about data and knowledge, but the human use of those for decision-making and action. Likewise, the 1996 policy said that information systems are "not only hardware and software but also associated personnel" (Department of Defense, 1996: 2) that exist within a larger "information environment" that includes "the aggregate of individuals, organizations, or systems that collect, process or disseminate information, also included is the information itself" (Department of Defense, 1996: enclosure 1–1). This represents a shift in understanding the space of information operations as being technological infrastructure to being complex socio-technical systems where humans are one of a number of heterogeneous elements (Law, 1987; Hughes, 1987).

This shift becomes even clearer in the 2001 iteration of the policy, which identifies decision-makers (i.e. humans) as "a primary focus of IO" (Department of Defense, 2001: 1). This, the policy said, would require paying attention not just to technological infrastructure but also to "human factors" such as "psychological, cultural, behavioral, and other human attributes that influence decision making, the flow of information, and the interpretation of information by individuals or groups at any level in a state or organization" (Department of Defense, 2001: enclosure 1–2). By 2016, the *Department of Defense Strategy for Operations in the Information Environment* defined the information environment (IE) as "a heterogeneous global environment where humans and automated systems observe, orient, decide, and act on data, information, and knowledge." The IE, it says, is composed of physical, informational, and cognitive dimensions. The last, the "human cognitive dimension," it says, is made up of "attitudes, beliefs, and perceptions," making it the "central object of operations in the IE." Cyberspace is smaller than this larger environment, "a global domain within the information environment" that encompasses systems and data but not the human and cognitive factors that are said to be most important (Department of Defense, 2016: 3).

Like cyber war, understandings of when and where information warfare/operations occur have expanded over time. Indeed, in their initial usages by Rona and Arquilla and Ronfeldt, respectively, cyber war and information warfare were largely the same thing – i.e. a mode of traditional, kinetic conflict marked by its central focus on the importance of information, communication, and associated technologies. Thus, cyber war and information war happened during times of war on traditional battlefields. Over time, this understanding has shifted, with the time and place of information operations broadening. Indeed, the shift in terminology from "warfare" to "operations" itself is indicative of this shift. Though the 1992 version of DODD 3600.1 focused on traditional military conflict, in 1996 and later versions, information operations are said to occur throughout the entire spectrum of conflict, where even "peacetime" is a time of conflict, and in a space that exceeds traditional, physical battlefields.

The centrality of the "human cognitive dimension" of an "information environment" larger than cyberspace has the effect of broadening the potential range of objects for attack and defense. Over time, U.S. DOD policy on information operations has included not just targeting the information and information systems of adversary militaries, but the attitudes, beliefs, perceptions, motives, reasoning, decision cycles, emotions, and will of the adversary population, groups within the population, as well as non-state groups. The wide spectrum of activities involved in this can be seen in the range of IW/O "core capabilities" and "related activities" identified in iterations of DODD 3600.1 over the years. These have included electronic warfare, operational security, information assurance, military deception, cyberspace operations, psychological operations/military information support operations, public affairs, civil-military affairs, (counter) intelligence, public diplomacy, influence activities, special technical operations, and even physical attack.

So far, these American understandings of IW/O seem to be a better fit for describing the broad, societally focused, cyber-enabled influence campaign that the Russians undertook in 2016 against the United States. However, the differences between the U.S. and Russian conceptions of IW/O begin to emerge when we look at the issue of defending against such attacks. Though U.S. policy contemplates a wide range of activities against the entirety of the adversary human cognitive dimension, as well as the expectation that adversaries will engage in such activities against the United States, discussion of what is to be defended, secured, or protected in these same documents is focused almost exclusively on DOD and military information, information systems, and decision-making. This is the case even in the most recent, 2016 DOD strategy for information operations, which is arguably the broadest yet in its conception of information operations (Department of Defense, 2016). In contrast, the most recent Russian doctrine on information security emphasizes the need to defend against information threats to "traditional moral and spiritual values" and to preserve the "cultural, historical, spiritual and moral values of the multi-ethnic people of the Russian Federation," several times warning of outsider attempts to use ICTs "to fester interethnic and social tensions, incite ethnic or religious hatred or hostility."[2]

So, despite the fact that there are a number of overlaps between the U.S. and Russian concepts of information warfare, the Russian definition is still slightly broader. What's more, if we understand the Russian operation in 2016, as well as similar, ongoing operations, as part of a larger campaign against not just the United States, but the EU and the West more generally, then even IW/O is still too narrow to encompass the full range of Russian actions – from hacking and social media manipulation, to political subversion, economic coercion, and even kinetic acts in the form of (barely) covert military operations and assassinations with the targeted use of WMDs (i.e. powerful, chemical nerve agents). Thus, some have proposed that concepts like hybrid or political warfare are more apt frameworks for understanding Russian actions. These will be the focus of the next two sections.

Hybrid warfare? They started it!

We saw at the beginning of this chapter that some have labeled ongoing Russian actions as hybrid warfare. This included some members of Congress, as well as former National Security Advisor Lt. Gen. H.R. McMaster, a noted military theorist in his own right. What's more, the term and concept saw its origins, in part, in a 2005 essay co-authored by former Secretary of Defense Gen. James Mattis (Mattis and Hoffman, 2005). Like cyber and information warfare, however, hybrid warfare has had differing meanings, some narrower and some broader, over time. This section explores the origins and evolution of these meanings. In its more recent, broader variant, hybrid warfare would include many of the kinds of activities Russia has undertaken in Estonia, Georgia, Ukraine, the EU, and the US since 2007. However, given the ongoing lack of clarity about the origins and definition of the term, especially as it relates to Russia, I argue that it is ultimately not the best term to describe Russian actions.

The term "hybrid warfare" seems to have originated in a Master's thesis project written by a Marine Corps officer at the Naval Postgraduate School in 2002 (Nemeth, 2002). In this initial usage, hybrid warfare was not about the kind of Russian activities of concern in recent years. Instead, the term "hybrid warfare" was used to describe the kind of warfare waged by "hybrid societies" that employed "hybrid militaries." Those societies and militaries were said to combine elements of modern and traditional societal and military norms and structures. The Chechen insurgency was highlighted as "a model for hybrid warfare" (Nemeth, 2002: v). The Chechen insurgency was described as using modern communication technologies, combined with "Soviet and Western doctrine" and "classical guerrilla strategies," to wage an "ancient form of warfare [i.e. guerrilla warfare] [at] new heights of efficiency" (Nemeth, 2002: 49).

Hybrid warfare gained much more attention in 2005 with the publication of an article authored by Lt. Gen. James Mattis and Frank Hoffman titled, "Future Warfare: The Rise of Hybrid Wars" (Mattis and Hoffman, 2005). Like Nemeth, and rooted in concerns over the challenges facing the United States in Iraq and Afghanistan at that time, Mattis and Hoffman framed hybrid warfare as an updated, high-tech version of insurgency or guerrilla warfare. Hoffman extended this framing in later work, including a 2007 report for the Potomac Institute for Policy Studies that framed hybrid warfare as a combination of conventional and irregular warfare with criminal activity. The 2006 war between Hizbullah and Israel was offered as a "prototype" for hybrid warfare (Hoffman, 2007). Later, Hoffman pointed to U.S. challenges in Iraq and Afghanistan as exemplary of the merging of conventional and irregular forms of warfare, high-intensity and long-term stability operations, in hybrid warfare. On a spectrum of conflict ranging from peacetime to major combat and global war, Hoffman placed hybrid warfare at the level of low- to mid-intensity conflict characteristic of terrorism, counterinsurgency, and "selective strike" (Hoffman, 2009: 6). That is, hybrid warfare is armed conflict (Hoffman, 2018: 32), rather than a form of conflict falling below the threshold of armed conflict as DOD policy documents often describe both cyberspace and information operations.

These definitions are in line with common understandings and usage of the term "hybrid warfare" in DOD more generally, at least up to the year 2010. As a result of various military leaders, including Gen. Mattis, using the term in Congressional testimony, members of Congress commissioned the Government Accountability Office (GAO) to conduct a study of the term's use and meaning within DOD. The report found that "DOD organizations . . . differed on their descriptions of hybrid warfare," with some using it to describe "a potent, complex variation of irregular warfare" and others merely to designate the expected increase in complexity of future conflicts (GAO, 2010: 2).

Though National Security Advisor McMaster's description of "hybrid warfare" discussed earlier would seem to encompass Russian operations against the U.S. and the West more generally, it is nonetheless out of line with the most common usages of that term in DOD. This adds an unnecessary level of confusion to the ongoing debate. To call what the Russians are doing "hybrid warfare" takes an existing term that is already ambiguous and redefines it in a way that makes it a much broader phenomenon than what is included in the more concrete definitions that already exist.

Perhaps more importantly, there is ongoing confusion around the term "hybrid warfare" specifically as it relates to Russia. There is a widespread misperception that this term is of Russian origin and reflective of official Russian doctrine. We see hints of that in Prof. Klaas' op-ed cited earlier, for example. However, as a number of Russia scholars have noted, this is incorrect. Russian officials' use of the term "hybrid warfare" is most often in reference to U.S. use of the term and reflects a misperception by the Russians that hybrid warfare is U.S. strategy. Thus, each side seems to think (incorrectly) that it is the other who created the concept of hybrid warfare and that this concept is reflective of official strategy and doctrine (Kofman and Rojansky, 2015; Adamsky, 2015; Galeotti, 2016). This hardly provides the basis for clear understanding.

Finally, the creation of new terms and concepts can have the negative effect of implying the existence of a new, perhaps unprecedented situation when none exists. The 2010 GAO report found that the reason DOD had no plans at that time to officially define and adopt the hybrid warfare terminology and concept was because it "does not consider it a new form of warfare" (GAO, 2010: 2). As of 2018, there was no official definition of this term in JP 1–02. Arquilla and Ronfeldt's "netwar," which I would argue also describes Russian actions fairly well, also suffers from this problem. Both terms imply that something that has a long history and basis of existing knowledge from which to draw for guidance is actually new and that technology is the most important driver of this change. In turn, this may imply the need to focus on technological fixes above all else. In the current situation, however, it has been our tendency to focus on technology, as evident in our narrow conception of cyberspace operations, which has been a major contributor to our failure to see the 2016 Russian operation coming. Adoption of terms and concepts that imply that the current situation is somehow new or unprecedented and that this is the result primarily of technological change are, therefore, counterproductive. As ADM Rogers said in his 2017 testimony cited earlier, though the cyber component was new, the overall contours of the Russian operation were not. Thus, in the next section we turn our attention to a historical analogy and concept of warfare that provides a better description of what we are seeing from Russia, as well as a deep reserve of prior experience and knowledge upon which to draw in crafting our responses.

Call of duty: Cold War black ops

Russian operations against the United States in 2016 are part of a broader, emerging conflict between Russia and the West. Snyder makes a strong case that this conflict is not just about isolated territorial, political, or economic disputes. Rather, it represents the emergence of a broader, cultural, and ideological conflict between Western, capitalist, liberal democracies and an increasingly nationalist, authoritarian, even fascist Russia working to spread its own model of society and governance and posing a heretofore under-appreciated threat to the current world order (Snyder, 2018). So far in this conflict, Russia has made use of cyberspace operations, but also propaganda, disinformation, subversion, economic coercion, support for guerrilla forces, assassinations, and even covert military operations. As we saw at the beginning of this chapter, various observers of this situation have been reminded of George F. Kennan's 1948 description of "political warfare."[3] While neither cyber nor information warfare quite fit the full scope of this emerging conflict, the concept of political warfare as articulated by Kennan and others at the start of the Cold War bears a striking resemblance to the situation we face today. If we are to understand and respond effectively to this new challenge, we may do well to spend less time inventing and debating new terms and concepts and instead hit the archives and the history books.

There is good reason to see parallels to our current situation in Kennan's May 1948 description of "political warfare." Kennan was not the sole developer of those ideas, however. His oft-cited memo on "The Inauguration of Organized Political Warfare" was but one contribution by a number of top U.S. national security officials to an ongoing discussion about the perceived need to conduct "black propaganda" and "psychological warfare" against the Soviets. Indeed, by the time Kennan wrote his memo, the National Security Council had already published two directives on the matter, one on overt and the other on covert propaganda, NSC 4 and NSC 4-A, respectively.[4] Examination of an archive of seventy-six documents, including correspondence and drafts of NSC directives, as well as existing historical research on these developments, sheds even more light on the broad sweep of what was included under "political warfare" during the Cold War.

Just as we worry today about the impacts of Internet and social media, this discussion about political warfare was occurring at the end of the second of two world wars where new technologies of mass communication had been used effectively by combatants on all sides. As early as the late nineteenth century, we can see increasing concern about the impacts of, and resultant competition among states to control, what has been called the "Victorian Internet" (Standage, 1998) – e.g. the telegraph – as well as the distribution of content from the state news agencies like the British Broadcast Corporation and others. The printing press, telegraph, telephone, radio, and manned flight with balloons and airplanes together resulted in a growing sense of both promise and peril about the use of the new technologies of mass communication by states and militaries during both peace and war (Examples of historical work on these topics include, Douglas, 1985; Simpson, 1994; Mattelart, 2000; MacDougall, 2006; Douglas, 2007; Butsch, 2008; Goodman, 2011; Auerbach, 2015).

Given the context, it is not surprising to see that for Kennan and others, political warfare was not perceived to be a new phenomenon even at that time. Various documents and memos in the series, including Kennan's, argued that other countries, including the British and the Soviets, had conducted political warfare for years, even before the war. Lenin was said to have made the idea even more potent, resulting in the U.S. need to counter ongoing Soviet political warfare. Kennan even acknowledged that the United States was actually already conducting political warfare, it just did not realize it and therefore was not acting in an organized and deliberate manner, reducing its effectiveness. He specifically identified the European Recovery Program (colloquially called the Marshall Plan) and U.S. support to the Western Union (a forerunner to both NATO and the EU) as examples of political warfare.[5]

Despite the increasingly intense competition with the Soviets and the great sense of urgency that comes across in this debate, Kennan and others were also clear that "political warfare" was not "war" as traditionally understood by Americans. Kennan faulted Americans for having a simplistic, black and white view of international relations that did not recognize the wide range of possible modes of conflict between peace and war. The situation with the Soviets was not war, but it certainly was not peace.[6] Others in the discussion seemed to understand and agree with this position. Secretary of State Marshall, for example, requested that the term "warfare" in psychological warfare be replaced with the word "operations," signaling his discomfort in framing the conflict as war.[7] Some military leaders expressed doubt about whether what they were discussing was really war and, as a result, expressed concern over the proper role of the military in helping to conduct covert political warfare, especially during peacetime. For example, Secretary of the Army Kenneth Royall at one point even suggested that the military should play no role at all.[8]

So, if political warfare was not war, what was it? Kennan defined it as "the employment of all the means at a nation's command, short of war, to achieve its national objectives." The range of tools, methods, and participants in such an effort was as broad as one might expect given such a broad definition. In Kennan's memo, they included both overt and covert, "white" and "black" methods. Pursuit of political alliances, the Marshall Plan, and overt propaganda were all "white" methods of political warfare. He also proposed the establishment of overtly funded and organized "liberation committees" of political refugees from the Soviet Union and Eastern Europe. But, much of Kennan's memo, and the debate overall, revolved around covert, "black" operations. For example, Kennan suggested covert operations, perhaps using American front companies, to undertake "Underground Activities behind the Iron Curtain." This would involve "establish[ing] contact with the various national underground representatives in free countries and through these intermediaries pass on assistance and guidance to the resistance movements behind the iron curtain." Another suggested "covert operation again utilizing private intermediaries" was "Support of Indigenous Anti-Communist Elements in Threatened Countries of the Free World." Finally, he suggested the need for "Preventive Direct Action in Free Countries." He wrote,

> This covert operation involves, for example, (1) control over anti-sabotage activities in the Venezuelan oil fields, (2) American sabotage of Near Eastern oil installations on the verge of Soviet capture, and (3) designation of key individuals threatened by the Kremlin who should be protected or removed elsewhere.[9]

Kennan's suggestions, along with those of others in the debate, shaped official policy on covert operations as set forth by NSC 10/2 in June 1948. Covert operations emerged through the course of this debate as a more generic term encompassing political, psychological, and economic warfare, as well as sabotage/anti-sabotage and other forms of "direct action." NSC 10/2 defined "covert operations" as

> all activities (except as noted herein) which are conducted or sponsored by this Government against hostile foreign states or groups or in support of friendly foreign states or groups but which are so planned and executed that any US Government responsibility for them is not evident to unauthorized persons and that if uncovered the US Government can plausibly disclaim any responsibility for them. Specifically, such operations shall include any covert activities related to: propaganda, economic warfare; preventive direct action, including sabotage, anti-sabotage, demolition and evacuation measures; subversion against hostile states, including assistance to underground resistance movements, guerrillas and refugee liberation groups, and support of indigenous anti-communist elements in threatened countries of the free world. Such operations shall not include armed conflict by recognized military forces, espionage, counter-espionage, and cover and deception for military operations.[10]

More specifically, in October 1948, the CIA's man in charge of covert operations, Assistant Director for Policy Coordination Frank Wisner, sent a memo to DCI Hillenkoetter outlining the initial program of "clandestine activities" he had developed based on NSC 10/2 and in preparation for the 1949 budget. The program included,

Functional Group I – Psychological Warfare

> Program A – Press (periodical and non-periodical)
> Program B – Radio
> Program C – Miscellaneous (direct mail, poison pen, rumors, etc.)

Functional Group II – Political Warfare

> Program A – Support of Resistance (Underground)
> Program B – Support of DP's and Refugees
> Program C – Support of anti-Communists in Free Countries
> Program D – Encouragement of Defection

Functional Group III – Economic Warfare

> Program A – Commodity operations (clandestine preclusive buying, market manipulation and black market operation)
> Program B – Fiscal operations (currency speculation, counterfeiting, etc.)

Functional Group IV – Preventive Direct Action

> Program A – Support of Guerrillas
> Program B – Sabotage, Countersabotage and Demolition
> Program C – Evacuation
> Program D – Stay-behind

Functional Group V – Miscellaneous

> Program A – Front Organization
> Program B – War Plans
> Program C – Administration
> Program D – Miscellaneous[11]

A 1948 Army document quoted by historian Christopher Simpson defined "psychological warfare" in more detail as including "overt (white), covert (black), and gray propaganda; subversion; sabotage; special operations; guerrilla warfare; espionage; [exploiting] political, cultural, economic, and racial pressures." Simpson reports that "special operations" was further defined in another Army document of the time as including "those activities against the enemy which are conducted by allied or friendly forces behind enemy lines. . . . [They] include psychological warfare (black), clandestine warfare, subversion, sabotage, and miscellaneous operations such as assassination, target capture and rescue of downed airmen." Such methods are effective, the documents said, "because they produce dissension, distrust, fear and hopelessness in the minds of the enemy" (Simpson, 1994: 12). Thus, Simpson sums up psychological warfare as being about the "exploitation of a target audience's cultural-psychological attributes and its communication system . . . the application of mass communication to modern social conflict: it focuses on the combined use of violence and more conventional forms of communication to achieve politicomilitary goals" (Simpson, 1994: 11).

Much of this early debate about covert political and psychological warfare revolved around who would be responsible for overseeing and carrying out such operations. Within the government, this authority was ultimately split between CIA, National Security Council, State Department, and DOD depending on whether operations were carried out during peacetime or wartime.[12] But, as Kennan's memo suggests, and as Simpson's research makes clear, much more of American society was mobilized in this effort. Beyond the government-organized liberation and refugee committees and participation by private businesses and other front organizations that Kennan suggested, Simpson demonstrates that the emergence of communication studies, a new scientific discipline focusing on issues of media effects, public opinion, and advertising, was aided to a significant degree by funding from the military, intelligence agencies, and State Department, as well as large private foundations, like Carnegie and Ford, that coordinated with the government in awarding funding for mass communication research (Simpson, 1994: 4). In short, U.S. covert political warfare was not merely a "whole-of-government" effort, but rather, a whole-of-society effort, involving government, business, philanthropic foundations, civic organizations, universities, Hollywood, and the news media.

Most of the targets of U.S. political warfare were overseas, with Europe the primary battleground in the early years of the Cold War. The primary documents examined here identify ongoing or planned operations in "friendly" nations like France, Germany, Italy, Greece, and Ireland. For example, officials worried about Soviet subversion of the Marshall Plan in France and considered whether making arrangements for a nation-wide showing of the new Hollywood movie, *Joan of Arc*, might help to buoy the French morale.[13] Operations in Italy went further and involved U.S. efforts to sway a national election in favor of the United States' preferred candidate (Mistry, 2006). Other documents discuss the use of U.S.-controlled territory in Germany as a base for radio broadcasts of propaganda into Eastern Europe and the Soviet Union,[14] as well as a plan (called Project Ultimate) to drop propaganda leaflets over Czechoslovakia using balloons (Cummings, 1999).[15] Of course, as we saw in Kennan's memo, covert physical actions like protection or sabotage of oil facilities in the Middle East or Venezuela were also on the list of possible operations. American covert political and psychological warfare operations would continue through the rest of the Cold War, from Korea to Vietnam to South America (Simpson, 1994; Solovey, 2001). Taking note of likely illegal State Department studies of domestic U.S. public opinion, Simpson claims that, at times, "the targets of U.S. psychological warfare were not only the 'enemy,' but also the people of the United States" (Simpson, 1994: 13).

Even for those unfamiliar with Cold War history, all of this should sound quite familiar to anyone paying attention to the conflict between Russia and the West today. The kinds of operations contemplated or carried out by the United States under the rubric of covert political warfare during the Cold War describe quite well the broad sweep of activities undertaken by Russia so far in the current conflict. This is true even if we focus only on the Russian operations against the United States, which have included use of new communication technologies for propaganda; covert support to internal political groups on the right and left, perhaps even to the President's own political campaign; election interference; exploitation of racial, political, and economic rifts in American society; and perhaps even support to secessionist movements in California and Texas. When we expand our scope beyond the United States, to include Russian actions in Estonia, Georgia, Ukraine, the European Union, and elsewhere, we can add coercive energy deals, support to guerrilla forces, covert direct military action, assassinations, and more to the list. Though the specific technologies and tactics of cyberspace and information operations are new (e.g. computer hacking, social media bots and trolls), the majority of what we are seeing thus far in the current conflict falls squarely within what the United States government has in the past officially called covert political, psychological, and economic warfare/operations.

Conclusion

This chapter began by noting the criticism of the dominant U.S. debate about cyber conflict in the wake of the 2016 Russian interference in the U.S. presidential election, as well as the profusion of terms and concepts deployed by officials and experts to describe Russian operations. It argued that the narrow conceptions of cyber conflict dominating U.S. national security discourse in the years leading up to the 2016 Russian operation left observers with inadequate language for talking about and, ultimately making sense of, what had happened. Of the options commonly debated, it argued that "political warfare," as defined in official U.S. policy during the early years of the Cold War, provides the best American "translation" for the broad sweep of Russian activities in an emerging conflict with the West that goes far beyond just interfering in the 2016 U.S. presidential election.

Nonetheless, as we know from the last chapter, no analogy or metaphor is perfect. That includes Cold War political warfare, meaning there is much work left to be done. If, at the end of the Cold War, we indulged in what Snyder calls "the politics of inevitability," entertaining fantasies of an "end of history" fueled by our own supposed technological prowess, he warns that we are now in danger of falling prey to the siren song of "the politics of eternity." In the former, the world is new, unprecedented, and on our side. In the latter, we are the innocent victims of a never-ending cycle of ancient conflicts and hatreds. Thus, in our current rhetoric of Russian cyber or information warfare, the Cold War is not just back, but perhaps it never ended after all – an example of letting our metaphors do our thinking for us once again. But there is also a tension in this rhetoric, in which the supposedly innocent United States has been attacked with cyber-IW that is both somehow unprecedented but also an age-old tactic that is just part of the "Russian nature." Russians will be Russians. Contradiction is also a hallmark of eternity thinking, Snyder says (Snyder, 2018). In the next and final chapter, we will review the book's argument thus far with an eye towards developing some broader lessons for how we might move beyond cyber-doom rhetoric, deploying language more effectively and avoiding the trap described by Professor Snyder.

Notes

1 Based on Google Scholar citation data.
2 A translation of the December 2016 document, "Doctrine of Information Security of the Russian Federation December," can be found on the Public Intelligence website at https://publicintelligence.net/ru-information-security-2016/.
3 Historical documents used in this section were accessed via an online archive provided by the U.S. State Department and are cited using endnotes containing the document number, title, and link to the document in the online archive. Kennan's famous memo, for example, is cited as 269. Policy Planning Staff Memorandum, https://history.state. gov/historicaldocuments/frus1945-50Intel/d269.
4 See 252. Memorandum From the Executive Secretary (Souers) to the Members of the National Security Council, https://history.state.gov/historicaldocuments/frus1945-50Intel/d252. Memorandum From the Executive Secretary (Souers) to the Members of the National Security Council, https://history.state.gov/historicaldocuments/frus1945-50Intel/d253.
5 269. Policy Planning Staff Memorandum, https://history.state.gov/historicaldocuments/frus1945-50Intel/d269.
6 269. Policy Planning Staff Memorandum, https://history.state.gov/historicaldocuments/frus1945-50Intel/d269.
7 250. Memorandum of Discussion at the 2d Meeting of the National Security Council, https://history.state.gov/historicaldocuments/frus1945-50Intel/d250.
8 250. Memorandum of Discussion at the 2d Meeting of the National Security Council, https://history.state.gov/historicaldocuments/frus1945-50Intel/d250. Memorandum for the President of Discussion at the 12th Meeting of the National Security Council, https://history.state.gov/historicaldocuments/frus1945-50Intel/d283. Memorandum for the President of Discussion at the 13th Meeting of the National Security Council, https://history.state.gov/historicaldocuments/frus1945-50Intel/d291.
9 269. Policy Planning Staff Memorandum, https://history.state.gov/historicaldocuments/frus1945-50Intel/d269.
10 292. National Security Council Directive on Office of Special Projects, https://history.state.gov/historicaldocuments/frus1945-50Intel/d292.
11 306. Memorandum From the Assistant Director for Policy Coordination (Wisner) to Director of Central Intelligence Hillenkoetter, https://history.state.gov/historicaldocuments/frus1945-50Intel/d306.
12 292. National Security Council Directive on Office of Special Projects, https://history.state.gov/historicaldocuments/frus1945-50Intel/d292.
13 307. Memorandum for the File, https://history.state.gov/historicaldocuments/frus1945-50Intel/d307.
14 261. Memorandum From the Chief of the Special Procedures Group (Cassady) to the Deputy Chief of the Special Procedures Group (Dulin), https://history.state.gov/historicaldocuments/frus1945-50Intel/d261. Memorandum From the Assistant Director for Policy Coordination (Wisner) to Director of Central Intelligence Hillenkoetter, https://history.state.gov/historicaldocuments/frus1945-50Intel/d306; and Director for Policy Coordination, Central Intelligence Agency (Wisner) to Members of His Staff, https://history.state.gov/historicaldocuments/frus1945-50Intel/d310.

15 293. Memorandum From Director of Central Intelligence Hillenkoetter to the Chief of Naval Operations (Denfeld), https://history.state.gov/historicaldocuments/frus1945-50Intel/d293. Memorandum From Commander Robert Jay Williams to the Chief of the Special Procedures Group (Cassady), https://history.state.gov/historicaldocuments/frus1945-50Intel/d296.

References

Adamsky, D. (2015) *Cross-Domain Coercion: The Current Russian Art of Strategy*, Paris: IFRI Security Studies Center.

Arquilla, J. and Ronfeldt, D. (1997) "Cyberwar Is Coming!" in Arquilla, J. and Ronfeldt, D. (eds) *In Athena's Camp: Preparing for Conflict in the Information Age*, Santa Monica, CA: RAND, pp. 24–60.

Auerbach, J. (2015) *Weapons of Democracy: Propaganda, Progressivism, and American Public Opinion*, Baltimore: Johns Hopkins University Press.

Brunner, E. and Dunn Cavelty, M. (2009) "The Formation of In-Formation By the US Military: Articulation and Enactment of Informatic Threat Imaginaries on the Immaterial Battlefield of Perception," *Cambridge Review of International Affairs*, 22, 4: 629–46. doi:10.1080/09557570903325454.

Butsch, R. (2008) *The Citizen Audience: Crowds, Publics, and Individuals*, New York: Routledge. doi:10.4324/9780203929032.

Chalfant, M. (2017) "Dem Senator: Russian Hacking May Have Been 'Act of War'," *The Hill*, 2 March 2017. Online. Available: <https://thehill.com/policy/cybersecurity/322002-dem-senator-we-should-determine-if-russian-election-hacking-was-act-of> (accessed 2 March 2017).

Clark, C. (2016) "'It Sucks to Be ISIL:' US Deploys 'Cyber Bombs,' Says DepSecDef," *Breaking Defense*, 12 April 2016. Online. Available: <https://breakingdefense.com/2016/04/it-sucks-to-be-isil-us-deploys-cyber-bombs-says-depsecdef/> (accessed 12 April 2016).

Clarke, R.A. and Knake, R. (2010) *Cyber War: The Next Threat to National Security and What to Do About it*, New York: HarperCollins.

Cummings, R.H. (1999) "Balloons Over East Europe: The Cold War Leaflet Campaign of Radio Free Europe," *The Falling Leaf*, Autumn, 166: 97–110.

Denning, D.E.R. (1999) *Information Warfare and Security*, New York: ACM Press.

Department of Defense. (1992) *Department of Defense Directive 3600.1, Information Warfare*, Washington, DC: Department of Defense.

———. (1996) *Department of Defense Directive 3600.1, Information Operations*, Washington, DC: Department of Defense.

———. (2001) *Department of Defense Directive 3600.1, Information Operations*, Washington, DC: Department of Defense.

———. (2011) *Department of Defense Strategy for Operating in Cyberspace*, 2011. Online. Available: <www.defense.gov/news/d20110714cyber.pdf> (accessed 2011).

———. (2015a) *Beyond the Build: Delivering Outcomes Through Cyberspace*, Washington, DC: Department of Defense.

———. (2015b) *DOD Cyber Strategy*, Washington, DC: Department of Defense.

———. (2016) *Department of Defense Strategy for Operations in the Information Environment*, Washington, DC: Department of Defense.

———. (2018a) *Achieve and Maintain Cyberspace Superiority: Command Vision for Us Cyber Command*, Washington, DC: Department of Defense.

———. (2018b) *Summary: Department of Defense Cyber Strategy*, Washington, DC: Department of Defense.

———. (2018c) *USCYBERCOM 2018 Cyberspace Strategy Symposium Proceedings*, Washington, DC: Department of Defense.

Douglas, S.J. (1985) "Technological Innovation and Organizational Change: The Navy's Adoption of Radio, 1899–1919," in *Military Enterprise and Technological Change: Perspectives on the American Experience*, Cambridge, MA: MIT Press, pp. 117–74.

———. (2007) "Early Radio," in Crowley, D. and Heyer, P. (eds) *Communication in History: Technology, Culture, Society*, Boston: Pearson, pp. 210–16. doi:10.4324/9781315664538.

Doyle, R. (1997) *On Beyond Living: Rhetorical Transformations of the Life Sciences*, Stanford, CA: Stanford University Press.

Galeotti, M. (2016) *Hybrid War or Gibridnaya Voina? Getting Russia's Non-Linear Military Challenge Right*, London: Mayak Intelligence.

———. (2018) "I'm Sorry for Creating the 'Derasimov Doctrine'," *Foreign Policy*, 5 March 2018. Online. Available: <https://foreignpolicy.com/2018/03/05/im-sorry-for-creating-the-gerasimov-doctrine/> (accessed 5 March 2018).

GAO. (2010) *Hybrid Warfare: Briefing to the Subcommittee on Terrorism, Unconventional Threats and Capabilities, Committee on Armed Services, House of Representatives*, Washington, DC: Government Accountability Office.

Goldman, A. (2018) "Justice Dept. Accuses Russians of Interfering in Midterm Elections," *New York Times*, 19 October 2018. Online. Available: <www.nytimes.com/2018/10/19/us/politics/russia-interference-midterm-elections.html> (accessed 19 October 2018).

Goodman, D. (2011) *Radio's Civic Ambition: American Broadcasting and Democracy in the 1930s*, New York: Oxford University Press. doi:10.1093/acprof:oso/9780195394085.001.0001.

Hasian, M., Lawson, S.T. and McFarlane, M. (2015) *The Rhetorical Invention of America's National Security State*, Lanham, MA: Lexington Books.

Hertling, M. and McKew, M.K. (2018) "Putin's Attack on the US Is Our Pearl Harbor," *Politico*, 16 July 2018. Online. Available: <www.politico.com/magazine/story/2018/07/16/putin-russia-trump-2016-pearl-harbor-219015> (accessed 16 July 2018).

Hoffman, F. (2007) *Conflict in the 21st Century: The Rise of Hybrid Wars*, Arlington, VA: Potomac Institute for Policy Studies

———. (2009) "Hybrid Threats: Reconceptualizing the Evolving Character of Modern Conflict," *Strategic Analysis*, April, 240: 1–8.

———. (2018) "Examining Complex Forms of Conflict: Gray Zone and Hybrid Challenges," *PRISM*, 7, 4: 30–47.

Hughes, T.P. (1987) "The Evolution of Large Technological Systems," in Bijker, W.E., Hughes, T.P. and Pinch, T.J. (eds) *The Social Construction of Technological Systems: New Directions in the Sociology and History of Technology*, Cambridge, MA: MIT Press, pp. 51–82.

Klaas, B. (2018) "Stop Calling It 'Meddling.' It's Actually Information Warfare," *The Washington Post*, 17 July 2018. Online. Available: <www.washingtonpost.com/news/democracy-post/wp/2018/07/17/stop-calling-it-meddling-its-actually-information-warfare/?utm_term=.5cdef2238d31> (accessed 17 July 2018).

Kofman, M. and Rojansky, M. (2015) "A Closer Look at Russia's 'Hybrid War'," *Kennan Cable*, April, 7: 1–8.

Landay, J. and Hosenball, M. (2018) "Russia, China, Iran Sought to Influence US 2018 Elections: US Spy Chief," *Reuters*, 21 December 2018. Online. Available: <www. reuters.com/article/us-usa-election-interference/russia-china-iran-sought-to-influence-u-s-2018-elections-u-s-spy-chief-idUSKCN1OK2FS> (accessed 21 December 2018).

Law, J. (1987) "Technology and Heterogenous Engineering: The Case of Portuguese Expansion," in Bijker, W.E., Hughes, T.P. and Pinch, T.J. (eds) *The Social Construction of Technological Systems: New Directions in the Sociology and History of Technology*, Cambridge, MA: MIT Press, pp. 111–34.

Lawson, S. (2011) "Articulation, Antagonism, and Intercalation in Western Military Imaginaries," *Security Dialogue*, 42, 1: 39–56. doi:10.1177/0967010610393775.

———. (2013) "The US Military's Social Media Civil War: Technology as Antagonism in Discourses of Information-Age Conflict," *Cambridge Review of International Affairs*, 27, 2: 226–45. doi:10.1080/09557571.2012.734787.

Lawson, S. and Middleton, M.K. (2019) "Cyber Pearl Harbor: Analogy, Fear, and the Framing of Cyber Security Threats in the United States, 1991–2016," *First Monday*, 24, 3. doi:10.5210/fm.v24i3.9623.

Libicki, M.C. (1995) *What Is Information Warfare?* Washington, DC: National Defense University. doi:10.21236/ada385640.

MacDougall, R. (2006) "The Wire Devils: Pulp Thrillers, the Telephone, and Action at a Distance in the Wiring of a Nation," *American Quarterly*, 58: 715–41. doi:10.1353/aq.2006.0062.

Makus, A. (1990) "Stuart Hall's Theory of Ideology: A Frame for Rhetorical Criticism," *Western Journal of Communication*, 54, 4: 495–514. doi:10.1080/10570319009374357.

Maness, R.C. and Jaitner, M. (2018) "There's More to Russia's Cyber Interference Than the Mueller Probe Suggests," *The Washington Post*, 12 March 2018. Online. Available: <www.washingtonpost.com/news/monkey-cage/wp/2018/03/12/theres-more-to-russias-cyber-meddling-than-the-mueller-probe-suggests/?utm_term=.e1fcd19ac6dd> (accessed 12 March 2018).

Mattelart, A. (2000) *Networking the World, 1794–2000*, Minneapolis, MN: University of Minnesota Press.

Mattis, J.N. and Hoffman, F. (2005) "Future Warfare: The Rise of Hybrid Wars.," *U.S. Naval Institute Proceedings*, 131, 11: 18–19.

McMaster, L.G.H.R. (2018) *Russian Aggression Is Strengthening Our Resolve*. Presentation at 100 Years of U.S.-Baltic Partnership: Reflecting on the Past and Looking to the Future, The Atlantic Council, Washington, DC.

Mistry, K. (2006) "The Case for Political Warfare: Strategy, Organization and US Involvement in the 1948 Italian Election," *Cold War History*, 6, 3: 301–29. doi:10.1080/14682740600795451.

Nemeth, W.J. (2002) "Future War and Chechnya : A Case for Hybrid Warfare," unpublished thesis, Naval Postgraduate School.

Pomerleau, M. (2018) "Where Do Information Operations Fit in the DOD Cyber Enterprise?" *Fifth Domain*, 26 July 2018. Online. Available: <www.fifthdomain.com/c2-comms/2018/07/26/where-do-information-operations-fit-in-the-dod-cyber-enterprise/> (accessed 26 July 2018).

Reuters. (2016) "Senator McCain Says Russia Must Pay Price for Hacking," *Reuters*, 30 December 2016. Online. Available: <www.reuters.com/article/us-usa-russia-cyber-mccain/senator-mccain-says-russia-must-pay-price-for-hacking-idUSKBN14J1LW> (accessed 30 December 2016).

Rona, T. (1976) *Weapon Systems and Information War*, Washington, DC: Office of the Secretary of Defense.

Samuelsohn, D. (2018) "Mueller Shifts Focus Back to Russian 'Information Warfare'," *Politico*, 16 February 2018. Online. Available: <www.politico.com/story/2018/02/16/mueller-indictment-russia-information-warfare-416378> (accessed 16 February 2018).

Schwartau, W. (1991) "Fighting Terminal Terrorism," *Computerworld*, 28 January 1991, 23.

———. (1994) *Information Warfare: Chaos on the Electronic Superhighway*, Emeryville, CA: Thunder's Mouth Press.

Simpson, C. (1994) *Science of Coercion: Communication Research and Psychological Warfare, 1945–1960*, New York: Oxford University Press.

Snyder, T. (2018) *The Road to Unfreedom: Russia, Europe, America*, New York: Tim Duggan Books.

Solovey, M. (2001) "Project Camelot and the 1960s Epistemological Revolution: Rethinking the Politics-Patronage-Social Science Nexus," *Social Studies of Science*, 31, 2: 171–206. doi:10.1177/030631270103100203.

Standage, T. (1998) *The Victorian Internet: The Remarkable Story of the Telegraph and the Nineteenth Century*, New York: Walker and Co.

Stein, G. (1995) "Information Warfare," *Airpower Journal*, Spring, 1995: 31–9.

Tomes, R.R. (2007) *U.S. Defense Strategy From Vietnam to Operation Iraqi Freedom: Military Innovation and the New American Way of War, 1973–2003*, London: Routledge. doi:10.4324/9780203968413.

Valeriano, B. and Jensen, B.M. (2017) "From Arms and Influence to Data and Manipulation: What Can Thomas Schelling Tell Us About Cyber Coercion?" *Lawfare Blog*, 16 March 2017. Online. Available: <www.lawfareblog.com/arms-and-influence-data-and-manipulation-what-can-thomas-schelling-tell-us-about-cyber-coercion> (accessed 16 March 2017).

Valeriano, B., Roff, H. and Lawson, S. (2016) "Dropping the Cyber Bomb? Spectacular Claims and Unremarkable Effects," *Net Politics*, 24 May 2016. Online. Available: <www.cfr.org/blog/dropping-cyber-bomb-spectacular-claims-and-unremarkable-effects> (accessed 24 May 2016).

Washington Post. (2017) "Full transcript: FBI Director James Comey testifies on Russian interference in 2016 election," *Washington Post*, 20 March 2017. Online. Available: <www.washingtonpost.com/news/post-politics/wp/2017/03/20/full-transcript-fbi-director-james-comey-testifies-on-russian-interference-in-2016-election/> (accessed 20 March 2017).

Watts, C. (2018) *Messing With the Enemy: Surviving in a Social Media World of Hackers, Terrorists, Russians, and Fake News*, New York: HarperCollins.

7   After action report and lessons learned

Introduction

Just as it seems not a day goes by without new reports of data breaches or other malicious activities in and through cyberspace, use of cyber-doom rhetoric is a persistent feature of the public discourse about cybersecurity in the United States. Even as the conclusion to this book is being written, new examples of cyber-doom rhetoric appear in the news and on political blogs almost daily. In June 2019, President Trump called off a planned airstrike against Iran in retaliation for its downing of a U.S. drone over the Persian Gulf. Instead, the President approved cyber operations against Iranian intelligence services and air defense systems (Nakashima, 2019; Barnes and Gibbons-Neff, 2019). In response, the right-wing political website *The Federalist* deployed the cyber Pearl Harbor metaphor and reference to Hurricane Maria to argue that "Trump's cyber attack on Iran was the right move." The author of the post warned that "the far more serious [cyber] threat" is attacks on critical infrastructure leading to a cyber Pearl Harbor that "could kill thousands of Americans very quickly" and for which "the impact of Hurricane Maria in Puerto Rico [is] a model of what might happen if we suffered such an attack." It was therefore important, the author claimed, for President Trump to send

> a clear message to Iran that [the United States] has a powerful cyber arsenal and is not afraid to use it. In fact, America might be able to inflict a cyber Pearl Harbor on Iran. Without dropping a single bomb, it may be able to unleash enormous death and destruction across wide swaths of Iranian territory.
>
> (Marcus, 2019)

In this one short example, we see various of the tactics of cyber-doom rhetoric, including appropriation of war and natural disaster, use of a real cyber event to focus our attention on what might happen instead of what actually happened (i.e. nothing close to a cyber Pearl Harbor or hurricane), as well as projection (i.e. the tacit implication that if we can carry out a cyber Pearl Harbor against others, perhaps it could happen to us).

On the same day as *The Federalist* piece, *Fox News* published its own cyber-doom warning. Opinion contributor Douglas MacKinnon, who is described as "a former White House and Pentagon official," asked, "Will you survive the coming blackout?" (MacKinnon, 2019). Amazingly, Mr. MacKinnon's essay deploys almost all of the tactics of cyber-doom rhetoric outlined in Chapter 2. Cyberattacks on the power grid, we are told, could result in "cataclysm" and "a nightmare scenario that so many fear." The word "cataclysm" is hyperlinked to a page on the *Fox News* website that aggregates articles on the topic of nuclear threats. In the middle of the article, we find a recommendation to another article, this one touting the possibility of a North Korean electro-magnetic pulse (EMP) attack and implying that such a threat is like "something out of a James Bond movie," an example of two different issues (i.e. cyber and EMP) being metaphorically linked together with the assistance of a reference to popular fiction (i.e. the "cuisinart effect" from Chapter 2). MacKinnon appropriated the June 2019 blackout in South America, which as we noted in Chapter 2 was not the result of a cyberattack, as "the largest red flag on this issue in years." He also appropriated the 2003 Northeast blackout and the numerous weather-related blackouts in the United States each year, which we saw in Chapter 4 did not result in anything close to cyber-doom effects, as evidence that "the clock is ticking" on cyberterrorist attacks against the grid. Such attacks, he says, are predetermined, "no longer seen as a question of 'If it will happen', but rather, 'When it will happen?'" "It truly is not a question of 'if', but of 'when'", we are warned – i.e. the tactic of determinism. To further reinforce the danger, he quotes from a Department of Homeland Security report on Russian infiltrations into critical infrastructure systems in 2018 that said, "they could have thrown switches" – i.e. the "it could have been worse," counterfactual tactic. Finally, like so many mass media fear appeals, MacKinnon's is long on promoting the supposed severity of the threat and the reader's susceptibility to it, but short on suggestions for what can be done to mitigate the threat. Indeed, he warns readers, "You and your family will be on your own. No one is going to ride to the rescue," and asks, "How will you survive?" His best suggestion: putting together a "two-week survival kit."

It is this persistence of cyber-doom rhetoric in the U.S. cybersecurity debate that served as the impetus for this book. As we saw in the very first chapter, cyber-doom scenarios like those expressed through the use of the cyber Pearl Harbor metaphor and analogy have been a feature of this debate from the very beginning. As the examples earlier illustrate, such scenarios often contemplate cyberattacks against critical infrastructure resulting in apocalyptic-levels of death, destruction, and collapse of society, economy, or even civilization. These scenarios and metaphors are but two of a collection of tactics in a broader cyber-doom rhetoric that persists despite growing recognition over the last decade – and especially since the 2016 Russian cyber interference in the U.S. presidential election – that such scenarios are not reflective of the real and growing cyber threats that we face and might even be a dangerous distraction. Thus, this book addressed the questions of why and how cyber-doom rhetoric persists, its characteristics, realism, and potential implications.

In answering these questions, this book has argued that cybersecurity and its reliance on doom rhetoric are key nodes in the articulation of the broader, U.S.

Information-Age security imaginary. Cyber-doom rhetoric operates as an appeal to fear that relies on various tactics, deployed individually or in combination, to focus attention on the worst possible outcomes. In doing so, it draws from and contributes to longstanding fears in American history and culture about the supposed role of infrastructure technologies, communication and transportation in particular, in creating and maintaining modern, democratic societies. Thus, the fears expressed in cyber-doom rhetoric are not so much about particular threat actors or even the immediate impacts of cyberattacks, but rather, about the supposed fragility of interconnected and interdependent societies, fears that long predate the advent of cyberspace. As illustrated most clearly by cyber-doom rhetoric's tactic of projection, the chief fear expressed in this rhetoric is ultimately of ourselves, our ability to control our creations, and our ability to cope if those efforts should fail. But these fears are largely misplaced. Examination of historical incidents of electrical blackouts, strategic bombing, terrorist attack, natural and human-caused disaster reveals that societies are more resilient than often assumed. What's more, reliance on expectations of social panic, paralysis, chaos, and collapse to frame our thinking about future threats can encourage a number of negative implications, including responses that are not only ineffective, but in some cases counterproductive and incompatible with democratic governance.

The next section of this final chapter will provide an "after action report" of sorts, reviewing the main arguments of the book. The final section will offer suggestions for how we might begin to think, speak, and act more effectively with respect to cybersecurity. It will encourage us to broaden our thinking and understanding of the nature of cybersecurity threats by drawing on a wider range of relevant knowledges and repertoire of metaphors and analogies. The goal of these efforts should be to promote greater social and technological resilience by encouraging decentralization and self-organization in our efforts to prevent and mitigate the effects of cyber threats.

After action report

We began the story of cyber-doom rhetoric's role in the U.S. cybersecurity debate by examining some key examples from popular and official discourse, such as media coverage of *Skyfall* and Secretary of Defense Leon Panetta's now infamous cyber Pearl Harbor speech, among others. We noted that despite numerous criticisms of such rhetoric over the years, including a growing realization by the U.S. intelligence community that cyber-doom is not the real threat in the years preceding the 2016 Russian cyber operations, nonetheless, such rhetoric remained a staple of public discourse. Since that time, and despite the fact that 2016 has served as an antagonism destabilizing the dominant U.S. cybersecurity discourse, still we see constant examples of cyber-doom rhetoric like the ones that began this chapter. The events of 2016 demonstrate that the critics of cyber-doom had been correct, but also that the staying power of cyber-doom rhetoric must be found in something other than its ability to accurately describe and predict current or future threats.

Reasons for cyber-doom rhetoric's persistence can be found, at least in part, in the fact that it is reflective of much deeper cultural and historical attitudes related to fear and technology in American society. First, the use of cyber-doom rhetoric is exemplary of the tendency among policymakers and advocates to believe that appeals to fear are effective tools for raising awareness and motivating a response to social problems. In general, cyber-doom rhetoric operates as an appeal to fear of "the uncontrollability of what could possibly happen in an uncertain world" (Walton, 2000: 14–15). As we saw in Chapter 2, the use of fear in public discourse is so prevalent that a number of scholars have documented what they call the emergence of a culture and politics of fear in the West generally, and the United States in particular. In this culture and politics, officials and experts serve as worst-case entrepreneurs, focusing public attention, often with the help of news and entertainment media, on a seemingly never-ending string of doomsday threats. In doing so, these scholars have observed, they draw from a standard menu of tactics. Advocates of greater cybersecurity are not immune to these broader social, cultural, and historical currents. They too are enmeshed in the wider culture and politics of fear that seems to dominate public discourse, thereby shaping and constraining the arguments and rhetorical resources available for making the case for better cybersecurity.

Cyber-doom rhetoric deploys many of the same tactics used by successful worst-case entrepreneurs in other areas of public life. First among these is what sociologist Barry Glassner called the "cuisinart effect." This involves the various mixings of fact and fiction examined in Chapter 2. This can include popular nonfiction like news media reporting on the results of "official fictions," like wargames and scenarios. It can involve official nonfiction like government reports or Congressional hearings deploying their own imagined scenarios or even referencing popular science fiction films and novels. In still other cases, news or educational media create their own fictional doomsday scenarios, sometimes with the help of former government officials or industry experts, to help "educate" the public.

Perhaps the second most common tactic that we encountered involves the appropriation of the fear and anxiety of current or historical non-cyber events, usually wars, terrorist attacks, or disasters, to call attention to cybersecurity threats. The ever-present cyber Pearl Harbor and cyber 9/11 metaphors and analogies are examples. But there are many others, such as analogies and metaphors to the Cold War and nuclear weapons, other terrorist attacks, hurricanes, tsunamis, oil spills, blackouts, and nuclear accidents. Almost any dramatic event, it seems, can be appropriated to warn of impending cyber-doom.

Next, researchers have long known that conflation and ambiguity can be deployed to bolster fear appeal arguments, especially in cases like cyber-doom where evidence of the threat might be lacking or audiences might be resistant to the message. As a number of observers have noted, officials and experts commonly deploy ambiguity in defining key terms and identifying just who might carry out a cyber-doom attack, against which targets, and with what potential responses, combined with conflating a number of very different types of cyber threats (e.g. crime, propaganda, infrastructure attacks) under one, monolithic category, all of which has the effect of inflating the likely threat of cyber-doom.

In still other cases, some cybersecurity advocates turn fact into fiction in an effort to draw our attention to the supposed threat of cyber-doom. In these cases, the impacts of actual cyberattacks can be exaggerated, as when an official or expert calls a cyberattack that resulted in no loss of life or destruction of property the fulfillment of cyber Pearl Harbor. In other cases, an actual cyber event might be highlighted not for the damage it actually caused, but to warn about what could have and might yet still happen in a future attack. This is the counterfactual, "it could have been (and likely will be) worse" story. Finally, we saw other cases in which our constant expectation of impending cyber-doom led officials, experts, and news media to see cyberattacks where none actually occurred, exaggerating the effects of otherwise normal accidents by framing them (at least initially) as the onset of the long anticipated cyber-doom.

This misreading of non-cyber events is enabled, at least in part, by the tactic of determinism. We see the potential onset of cyber-doom where none exists because we have been led to believe that its onset is predetermined, that it is inevitable, that it is, as MacKinnon said at the start of this chapter, a matter of when not if.

Finally, cyber-doom rhetoric often relies on the tactic of projection, which involves using specific malicious incidents, activities, or capabilities contemplated, carried out, or caused by the United States as a means of raising fear of the cyber threats that others might pose to the United States. Like cyber Pearl Harbor, this tactic too has been a staple of the U.S. cybersecurity debate from the beginning. Taken together, then, cyber-doom rhetoric operates as a fear appeal that employs a collection of rhetorical tactics that reinforce the twin, master metaphors of war and disaster to frame our thinking about cybersecurity, always seeking to focus our attention not on what is, but instead on the supposed inevitability of the worst-case scenarios to come.

But cyber-doom rhetoric is not just exemplary of and enabled by the contemporary culture and politics of fear and the various tactics used to create and maintain it. Its persistence is also aided by the fact that it taps into and reflects even more deep-seated beliefs in American history about the supposed role of technology in the creation and maintenance of modern, democratic societies and concomitant fears of what might happen if that technology were to fail or get out of control. Historians have documented that even from the earliest days of America's founding, technology was viewed as a positive, even essential force for good. American founders like John Adams and Thomas Jefferson, for example, believed that technologies of transportation and communication, what we would today call "critical infrastructure," would be essential for holding the nation together across vast geographical distances. Over time, many came to believe that the nation's success would not just be determined by its technology, but that technology was the driving force of all human history and progress.

Over the course of the late nineteenth and twentieth centuries, however, we see a creeping pessimism about the role of technology in society and our ability to control our creations. This was the result of negative experiences like world wars

and the use of frightening new weapons. But it was also the result of the changing character of technology over the same period. Where technology once had
referred to standalone artifacts that one could readily see, touch, and presumably
control, the Victorian Age saw the rise of large technological systems like transcontinental railroads and telegraph networks. These were systems that one could
not readily apprehend in their entirety. Indeed, as we saw in Chapter 3, many of
the fears and anxieties expressed at the turn of the twentieth century about telegraph, telephone, radio, and electrification are almost identical to contemporary
concerns about cybersecurity. This is not surprising as the Internet and cyberspace
more generally are the best examples yet of the kinds of large socio-technical
systems that began to cause such anxiety over the last century.

These fears of technology-out-of-control have exerted an important influence
on Western military thought during the same period. From theorizing industrial
age, mechanized warfare in the early twentieth century to Information-Age, computerized warfare at the turn of the twenty-first century, military strategists and
civilian defense intellectuals have worried that new technologies simultaneously
promise new capabilities, but also the prospect for doom if we fail to keep pace.
Military thinkers often take a deterministic and ambivalent view of new technology – whether airplanes or computers – and its implications for the future
of warfare. In both cases, modern societies are depicted as newly and uniquely
dependent on complex technologies that are not well understood, offering the
opportunity to strike a knockout blow against an adversary or, more concerning,
for the adversary to deliver such a blow against us. In both cases, we also see a
tendency for ambivalence towards the new technologies not to lead to rejection of
their development and use, but rather, to intensified adoption in an effort to keep
pace and remain in sync with socio-economic currents supposedly driven by technological forces out of our control and necessitating our response. But we also
see the use of projection to justify continued adoption of new technologies, with
military strategists and civilian defense intellectuals sometimes incorrectly seeing in our adversaries the capabilities and actions that we desire or have already
developed and deployed. Finally, we see the increasing reliance over the course of
the last century on fictions of various kinds to help military strategists and civilian
defense intellectuals think through the implications of new technologies for the
future of warfare, from the use of literature to computer gaming and simulation.

But we also know that in many cases the expectations of technologically
enabled quick victories or doomsday scenarios contemplated by these thinkers
turned out to be off the mark. Indeed, whether as the result of strategic bombing, terrorist attack, or disaster – i.e. the most common metaphors and analogies
used to frame our discussion of cybersecurity today – infrastructure failures have
not tended to result in the kinds of panic, paralysis, and chaos often predicted in
cyber-doom scenarios. Historians have documented that the predictions of quick
victory from the air did not pan out in either of the world wars. Such predictions, they said, should have been suspect from the start as real life examples of
blackouts, for example, demonstrated that societies were not as sensitive to infrastructure failures as the military planners assumed. Instead, however, planners

continued to rely on hypothetical scenarios, assuming that infrastructure failures caused by intentional attack would have different outcomes than those caused on accident. When this did not turn out to be the case, they redoubled their efforts, bombing even more intensely, up to and including firebombing of entire cities and the dropping of two atomic bombs. And yet, as historians have noted, cities did not die and populations proved more resilient than expected, even defiant in the face of aerial assaults that seemed only to prove the necessity of unconditional victory against a seemingly barbaric enemy.

Even still, Cold War planners assumed that the advent of atomic weapons would certainly deliver the effects that strategic bombardment with conventional munitions had failed to achieve. This spawned the new field of disaster sociology, the goal of which was to study the way populations respond in disaster situations and apply that knowledge to civil defense efforts. But the sociologists have consistently returned results at odds with the assumptions that underlay the founding of their field: societies are actually quite resilient in the face of disaster, that is, if government does not take actions that are counterproductive, which is too often the case. The sociologists continued to demonstrate these findings into the twenty-first century, where terrorist attacks like 9/11, mass blackouts like the one in 2003, and natural disasters like Hurricane Katrina did not result in the kinds of social and economic panic, paralysis, chaos, and collapse often predicted by officials or depicted in news and entertainment media.

It should come as no surprise, then, that even the most significant examples of state-sponsored "cyber war" to date have not approached the kinds of effects contemplated in cyber-doom rhetoric. So far, at least, critical infrastructures have not been the primary targets of the vast majority of cyberattacks, including in the most often cited cases of state-sponsored "cyber war." On the occasions that critical infrastructures have been targeted, the effects of those attacks have been quite limited. In each instance, however, initial reactions include expression of the idea that the attack is the fulfillment of the long-predicted cyber-doom, followed by more sober assessments of the situation, then often followed with a form of the "it could have been (and likely will be) worse" rendering of the event. We see this pattern in the cyberattacks against Estonia in 2007, Georgia in 2008, Iran in 2010, and Ukraine in 2015–2017. In fact, though each of these cases initially drew reactions proclaiming the long-awaited onset of cyber war, over the course of a decade they came instead to be seen as examples of the limitations of cyber operations to achieve decisive, strategic effects and as examples that the nature of international cyber conflict was turning out to be quite different than what decades of cyber-doom rhetoric had led us to expect.

These repeated instances of failed predictions and initial misreading of events are perhaps the best evidence of the potentially negative implications of over reliance on cyber-doom rhetoric. Language and rhetoric, including metaphor and analogy, are not just a matter of literary flourish or decoration. Rather, as cognitive scientists have demonstrated, language and metaphor are central to the way we think, enabling and constraining certain ways of seeing, understanding, and therefore acting in the world. Scholars as far back as the ancient Greeks have known that appeals to fear of the kind often found in cyber-doom rhetoric are prone to failure if not deployed carefully. In particular, the threat and response components of a fear appeal must be balanced, seeking not only to scare audiences about the potential severity of a threat and their susceptibility to it, but offering responses that the audience will believe they can take and will be effective in mitigating the threat. Too often, policymakers and advocates focus on the threat's severity and the audience's susceptibility to it, which can have the effect of encouraging either fatalistic under-reactions or costly and counterproductive reactions. As noted in Chapter 2, many mass media depictions of cyberattack fall prey to precisely this pattern of over emphasizing the severity of cyber threats while ignoring potential responses or encouraging centralized, militarized responses.

Metaphors and analogies also play a key role in our thinking and responses to social problems of various kinds, including cybersecurity threats. Like fear appeals, however, war and disaster metaphors in particular can easily lead us astray. The war/disaster framing can encourage a number of distorted ways of thinking and ineffective or counterproductive responses. This can include inflation of the severity of the threats we face, but also distraction from other threats that may be more realistic but less amenable to martial framings. In other cases, we can have the tendency to mistake our metaphors for reality, what one scholar has called "metaphorical idolatry." Next, war/disaster metaphors and the severity of consequences that they imply can cause us to neglect the probability that such scenarios will actually occur, encouraging us to treat worst cases as inevitable or to treat any level of risk, no matter how small, as warranting a preemptive response. In turn, this way of thinking can cause us to ignore our responsibility for actions we take, seeing them as necessary and therefore justified, or externalizing the blame for our actions and their implications. Finally, war/disaster framings and their distorted ways of thinking can encourage ineffective, costly, and counterproductive responses, including militarization and anti-democratic forms of governance.

We can see numerous indicators that the primary U.S. responses to cybersecurity threats are falling prey to the negative implications of cyber-doom rhetoric. We can see numerous examples over decades where officials and experts warn of impending cyber-doom while the real world continues to defy their predictions. We have also observed the tendency to let our metaphors do our thinking, too often forgetting that cyberspace is not really a place or a domain, that cyber weapons are in many ways not like weapons, and that deterrence is perhaps not the same or even applicable to cyber conflict in the way it was to nuclear warfare. In fact, much of what gets called cyber "war" is not war at all as typically understood.

And yet, the most significant U.S. response to cybersecurity threats to date has been the creation of a military Cyber Command. In some cases, the actions taken by this command or its progenitor, the National Security Agency – a military intelligence organization – have either been inapt and ineffective or even counterproductive. It is not at all clear that a military command or intelligence organization is the appropriate means of responding to cyber crime, intellectual property theft, and propaganda, for example. In other cases, actions taken by NSA or Cyber Command would seem to undermine cybersecurity or increase the risks of ushering in precisely the kinds of cyber-doom scenarios we seek to avoid. This is the case when these organizations either ignore or even seek to create and exploit security flaws, which, in some cases, have ended up being discovered and used by criminals or nation-state adversaries against the United States. In still other cases, NSA or Cyber Command has engaged in precisely the kinds of infrastructure hacking so many have worried could result in cyber-doom scenarios, thereby increasing the risk that cyber-doom becomes a reality after all.

Perhaps the most significant negative implication to our obsession with cyber-doom was the failure to see the 2016 Russian cyber operations coming. While cyber-doom rhetoric is certainly not entirely to blame for that failure, as we saw in Chapter 1, a number of experts have claimed that it at least played a part. Whatever the case, the 2016 Russian cyber-enabled information operations during the U.S. presidential election served as an antagonism that destabilized the dominant U.S. cybersecurity discourse by undermining the position of cyber-doom rhetoric. We can observe the effects of this by noting the shift in language that has occurred since 2016, in particular the sudden emergence of new terms and concepts, re-emergence of older terms and concepts, and general confusion and debate about just what to call the Russian operations and what they mean for the future of cyber conflict.

The terms and concepts dominating the current debate about Russia and the future of cyber conflict include cyber warfare, of course, but also information warfare, hybrid warfare, and political warfare. Of these, the last chapter argued that political warfare is the most appropriate "translation" based in the United States' own history and experience. As we have discussed throughout this book, cyber warfare comes with the baggage of cyber-doom rhetoric, which some have claimed is partly to blame for failing to see the Russian operations coming. What's more, historically, our discussion of cyber warfare in the United States has been focused almost entirely on technical features of cyberspace, largely ignoring the social and cultural components that have emerged as central to the kinds of operations Russia carried out in 2016 and after. Likewise, in the U.S. context, information warfare has had a shifting and ambiguous meaning over time, at one moment focusing on technical matters, making it a synonym of cyber warfare, at other times focusing more broadly but still mainly concerned with attacking and defending military networks, personnel, and data, rather than the wider populace. Similarly, there is confusion over the meaning and scope of hybrid warfare. Some, like its American originators, see it as a form of armed conflict combining traditional and irregular tactics. Others, like the United States' NATO allies, often have a much broader conception of hybrid warfare that includes many activities that fall below the threshold of armed conflict. And then there is the matter of confusion over just who originated the term and concept, the Americans or the Russians. Some Americans claim that the Russians invented it and that it is their official doctrine while many Russia experts point out that the Russians are reacting to the American use of the term and concept and assuming that it reflects official American strategy. As a result of these limitations and confusions, the last chapter argued that, ironically enough, the Cold War might yet provide the most appropriate analogy for thinking about the current state of cyber conflict. That is, the use of new technologies of mass communication for political warfare between peer competitors might provide a better historical starting point for thinking about the current situation than the alternatives.

Beyond cyber-doom

None of the preceding discussion should be read as suggesting that we should not take cybersecurity seriously, that we should not take measures to secure our critical infrastructures, or that we should not prepare to mitigate the effects of a cyberattack on critical infrastructure should it occur. Rather, this book should be read as suggesting that taking cybersecurity seriously requires that we re-evaluate the assumptions upon which public debate and policymaking proceed, that we can only make effective policy if we begin with a realistic assessment of current and likely future threats. To do that, we must continue to challenge and ultimately move beyond the rhetoric of cyber-doom that has persisted in the U.S. cybersecurity debate for more than two decades.

The first step to doing that is to focus our attention on reality over fiction and fantasy. As we noted in Chapter 2, sociologists studying our culture and politics of fear observe that policymakers and advocates too often replace facts with scenarios such that the line between fact and fiction becomes almost indistinguishable. We saw a number of examples of this in the history of the U.S. cybersecurity debate. To be clear, just as we cannot and should not avoid the use of metaphors, wargames, scenarios, and simulations can be valuable tools for thinking through potential futures of cybersecurity and conflict. But we have also seen that many uses of cyber-doom scenarios serve no purpose other than promoting fear, which we know can and does often backfire.

Thus, our public policy debate over the formulation and evaluation of cybersecurity policy and strategy needs to be guided whenever possible by empirical research and rely less on hypothetical scenarios. In the case of early airpower theory, we saw in Chapter 4 that reliance upon unchallenged assumptions and hypothetical scenarios in the face of contradictory empirical evidence had disastrous results. By relying too heavily on hypothetical, cyber-doom scenarios, current cybersecurity planning is open to the same criticism that has been leveled against contemporary disaster planning, which is that it is "organized to deal with predicted vulnerabilities rather than to mobilize social capital to deal with actual threats" (Dynes, 2006). Fortunately, we have seen a growing body of empirical work on cybersecurity and conflict in recent years, in particular from political scientists, that can help to improve the situation (Lindsay, 2013, 2015; Valeriano and Maness, 2015; Valeriano et al., 2018; Brantly, 2016; Kostyuk and Zhukov, 2019).

When we focus on reality, we see that cyber conflict is not just turning out to be different than we had expected, but much broader and even more complex. This reality demands that we think more broadly about cybersecurity and conflict, recognizing that the challenges we face are not merely matters of technical or military concern. Cybersecurity and conflict today contain deeply imbricated social, cultural, and historical dimensions that we can no longer ignore.

Taking the breadth and complexity of real cybersecurity challenges seriously will require hearing from a wider diversity of voices and drawing from a broader range of knowledges and metaphors to shape the way we think, speak, and act. The growing body of work on cybersecurity and conflict in political science over the last ten years is a good start. For too long, current or former government officials and IT industry executives or consultants dominated the public debate in the United States (Lawson, 2010). Of course, these experts should have a voice and a seat at the table. But we must hear from many more experts from a diversity of backgrounds and fields of research. One goal of this book has been to counter the tendency towards "you never knowism" and possibilistic thinking that we encountered in Chapters 2 and 5 by demonstrating the value of research conducted in the humanities and social sciences, in particular the history of technology, military history, disaster sociology, rhetoric, communication studies, and psychology for the analysis of cybersecurity and conflict.

There has been a shift in this direction happening within the U.S. military for quite some time that is only just beginning to impact thinking about cybersecurity and conflict. The highly influential creator of the observation-orientation-decision-action (OODA) loop theory and promoter of maneuver warfare that we encountered in Chapter 3, retired Air Force Colonel John Boyd, also argued that success in conflict of any kind requires "selecting information from a variety of sources or channels" for creating more appropriate "mental images" to guide our decisions and actions (Boyd, 1987b: 49). He advocated a "kind of thinking" that looked to multiple domains of knowledge for insights, from history to natural science, and even Eastern philosophy (Boyd, 1987a: 2). Boyd's work and the "kind of thinking" he advocated have been at the heart of the design movement within the U.S. and allied militaries over the last decade. These military professionals have promoted drawing from diverse sets of knowledge, as well as understanding the role of language and metaphor in problem setting and framing, as keys to success in increasingly complex, multi-domain conflicts (Examples of military design thinking include, Cardon and Leonard, 2010; Elkus, 2010; Banach and Ryan, 2009; Wass de Czege, 2009; Schmitt, 2006). For example, the work of Donald Schön on generative metaphor and framing that we encountered in Chapter 5 has been particularly influential to these military thinkers, along with a wide range of work from philosophy to management science not traditionally seen as relevant to military concerns.

Applying this kind of thinking and its understanding of the role of language and metaphor to cybersecurity and conflict would mean casting our net more broadly to other domains of human experience for insight. Doing so in Chapters 3 and 4 revealed that contemporary fears related to cybersecurity are exemplary of long-standing fears and assumptions about the relationship between technology and society that could very well be distorting our thinking. Continuing to cast our net broadly may reveal other deep-seated, cultural fears, anxieties, and assumptions at work in the U.S. cybersecurity debate. For example, beyond noting the important role that ideas about communication and transportation technology have played in American history, media historian James Carey also detailed how these beliefs were accompanied by even more fundamental assumptions about the nature of information and communication. He argued that Americans have historically held a rather simplistic, "transmission view" of communication that focuses on the use of technological infrastructures for the sending and receiving of bits and bytes of information. He advocated, instead, for a more nuanced, "ritual view" of communication as a complex process of cultural formation that goes far beyond the mere sending and receiving of information and messages (Carey, 1989). Other scholars have begun to note the role that these competing understandings of information and communication have played in the U.S. experience with Cold War political warfare and more recent theorizations of information warfare and the emerging role of social media in complex, multi-domain conflicts (Simpson, 1994; Brunner and Dunn Cavelty, 2009; Lawson, 2013). As our understanding of the nature of cybersecurity and conflict begins to recognize the complex challenges of cyber-enabled information operations, more work is needed to unpack the way that these assumptions about the very nature of information and communication may be shaping our thinking, and with what impacts.

We have begun to see some progress on these issues within the U.S. military cyber debate. For example, in 2009, U.S. Strategic Command, then the parent organization of USCYBERCOM, released *The Cyber Warfare Lexicon*, which began with a series of epigraphs that emphasized our inability to think clearly and effectively about a subject without having the right language (USSTRATCOM, 2009). To help produce better language to think with, in 2012, USCYBERCOM launched the Cyber Analogies Project at the Naval Postgraduate School with the mission "to assist U.S. Cyber Command in identifying and developing relevant historical, economic, and other useful metaphors that could be used to enrich the discourse about cyber strategy, doctrine, and policy" (Goldman and Arquilla, 2014: 1). Ultimately, the report argued that appropriate "analogies, metaphors, and parables" are necessary to effectively understand and speak about cybersecurity and conflict (Goldman and Arquilla, 2014: 5–6). This is a good start, but more work is needed to continue exploring the range of possible metaphors and analogies from various fields.

To mitigate the potentially negative implications of over reliance on martial metaphors, Mongoven advised the deliberate pursuit of "a different symbolic framing with reflective metaphorical diversification . . . highlighting what military metaphors overemphasize or obscure, as well as challenging them with alternative metaphors" (Mongoven, 2006: 414–15). Doing so, von Ghyczy advised, requires flipping the way we typically use metaphors on its head. In addition to confusing metaphors and models, which we discussed in Chapter 5, he said that we often rely on the use of "rhetorical metaphors," in which we seek understanding of the new in terms of the already known and familiar. Opening up our thinking in innovative and creative ways, however, requires the use of what he calls "cognitive metaphors," which work in the opposite way to rhetorical metaphors. He explained,

A cognitive metaphor juxtaposes two seemingly unrelated domains of reality. Whereas rhetorical metaphors use something familiar to the audience (for example, the infectious virus, which passes from person to person) to shed light on something less familiar (a new form of marketing that uses e-mail to spread a message), cognitive metaphors often work the other way around. They may use something relatively unfamiliar (for example, evolutionary biology) to spark creative thinking about something familiar (business strategy).

(von Ghyczy, 2003: 89)

The goal, he cautions, is not to become an expert in the unfamiliar domain, it is to learn enough about a seemingly unrelated and unfamiliar domain to allow us fresh insights on the problems with which we are contending. In short, adopting Boyd's "way of thinking" and drawing from a broader range of potentially relevant knowledges can help to develop the wider, more diverse repertoire of metaphors for thinking and speaking about cybersecurity and conflict. There is no shortage of options from which to choose, with scholars and practitioners over the years recommending deeper exploration of a number of alternative metaphors, some still martial in nature, but many drawn from quite different fields. On the martial side, these include metaphors to political, economic, guerrilla, air, or biological warfare (Liles, 2010; Lawson, 2012; Goldman and Arquilla, 2014). But there are many more non-martial metaphors that we might fruitfully explore, including ecosystems, immune systems, epidemiology and public health, complex adaptive systems, mathematical topology, and many more (Libicki, 1997; Charney, 2010; JASON, 2010; Department of Homeland Security, 2011; Lapointe, 2011; Axelrod, 2014; Goldman and Arquilla, 2014; Finn, 2016).

No metaphor is perfect and, in fact, if used appropriately, von Ghyczy tells us, these imperfections are a source of creativity and innovation (von Ghyczy, 2003). Thus, given the tendency towards metaphorical idolatry mentioned in Chapter 5, we must remain on guard not to let any of these metaphors do our thinking for us. This includes Cold War political warfare. Just as the war/disaster metaphors at the heart of cyber-doom rhetoric have serious limitations, the alternative metaphors highlighted earlier have their own limitations. Any of the alternative, martial metaphors may still imply a need for militarized responses. But even biological or life sciences-based metaphors can be potentially problematic. They can have a tendency to naturalize social phenomenon and, thereby, encourage deterministic ways of thinking. We see this in various attempts to understand social realities in biological terms, including structural functionalism, sociobiology, and the "selfish gene." Life science metaphors are also not immune to militaristic entailments. For example, industrial and Information-Age theories of warfare drew variously from eugenics, evolutionary biology, crowd psychology, complexity theory, and more (Lawson, 2014). Finally, the life sciences have themselves been inflected with militaristic metaphors. Donna Haraway, for example, has documented the way the immune system has been described by scientists as a command and control system for defending the body against foreign invaders (Haraway, 1991; Libicki's application of the immune system metaphor to information warfare is an example, see Libicki, 1997). In short, life science-based metaphors are no guaranteed solution to the problem of militarization of cybersecurity.

Likewise, political warfare, the metaphor and analogy we identified as potentially most appropriate in the last chapter, comes with its own set of important limitations. History is never entirely new and unprecedented, nor entirely the same, and certainly not inevitable. The application of history to current affairs, therefore, is always an exercise in analogical or metaphorical reasoning in which we attempt to sort out sameness and difference between two situations that will never entirely match up. To do that effectively, we must be honest with and about ourselves. As Sun Tzu reminded us, we must know the truth of the enemy, but also of ourselves. One step in that direction is to recognize that we have been to some version of "here" before, that we have engaged in political warfare ourselves. What the Russians are doing now sounds an awful lot like the political, economic, and psychological warfare that Americans engaged in during the Cold War.

And yet, none of this means our current situation is merely "the same," nor that Russian actions are of no concern, and certainly not that the United States "had it coming," so to speak. Russia is not the Soviet Union, either in power or ideology. But, neither is the United States the same as it was in the late 1940s and 1950s, or even the 1980s. The level of wealth inequality is much higher and our commitment to public welfare much lower than it was during the first thirty-five years of the Cold War, a situation exploited by the Russians in their efforts to undermine American political institutions. Of course, the technology too is different, seemingly providing newfound capabilities to implement old strategies with a new degree of effectiveness.

There will be no easy answers, just as there were none during the Cold War. The United States made a lot of mistakes. Though some on the far right and far left today use McCarthyism as a form of "whataboutism" to downplay or even deny Russian interference in the 2016 election, that historical episode nonetheless should provide a cautionary tale of what can happen when we take our fears of foreign political warfare out on ourselves. Then, as now, if we undermine the core values and liberties that we believe make us "exceptional," then we have allowed the adversary to win. Ultimately, our challenge will be to understand the current situation for what it is, which is political warfare. We must understand that we have been here before and use that as a foundation for articulating precisely what is the same and what is different about our current situation. We will then stand a far better chance of responding effectively, as we have in the past, while hopefully avoiding our costly past mistakes.

At a more abstract level, thinking seriously about metaphor reminds us to pay attention to the play of sameness and difference. Just as contemporary cyber conflict looks a lot like Cold War political warfare without being exactly the same, we must pay closer attention to the similarities and differences among the various cybersecurity threats that we face. This requires rejecting the monolith thinking and conflation described in Chapter 2 by recognizing that there is not one cyber threat but many. As James Lewis has argued, "Pronouncements that we are in a cyber war or face cyberterror conflate problems and make effective response more difficult." Instead, he advocates that we disaggregate the different types of cyber threats – including cyberspace-enabled economic espionage, political and military espionage, crime, and cyber war or cyberterrorism – so that each threat can be addressed in the most appropriate and effective manner (Lewis, 2010: 1). There is no one-size-fits-all solution. Myriam Dunn Cavelty goes even further, urging us to take the complexity of contemporary cybersecurity problems seriously, not by reducing all challenges to "cyber war" and thinking in terms of "hysterical doomsday scenarios," but instead by focusing "on a far broader range of potentially dangerous occurrences involving cyber-means and targets, including failure due to human error, technical problems, and market failure apart from malicious attacks" (Dunn Cavelty, 2008: 144). For effective response, complex problems like contemporary cybersecurity challenges require us first and foremost to acknowledge their complexity and to work towards the clearest, most precise definitions possible, even when absolute clarity and precision are unattainable.

Nonetheless, one thing we learn from Russian election interference is that the small things matter. The vulnerabilities that allow criminals to extort or steal identities are the same ones that allow state actors to engage in election interference. They are the same ones that could be exploited for critical infrastructure attacks. Thus, a broader range of participants must be empowered to act, including individuals. Not only is there something effective that average people can do to help mitigate such threats – e.g. learning and applying basic principles of "cyber hygiene" (Finn, 2016) – but we must get their help. The military and intelligence agencies cannot and should not solve this problem alone.

Participation by a wider range of actors, from individuals to private organizations and civilian agencies, will be necessary to promoting greater social and technical resilience. While it is unclear whether the possession of an offensive cyber war capability will deter potential attackers, it is clear that more resilient technological and social systems are a benefit in any case. They can help to mitigate the effects of a cyberattack should it occur and might even help to deter cyberattacks by providing a would-be attacker with fewer valuable and vulnerable targets (Lewis, 2006; Nye, 2010: 189, 191). Promoting resilience, however, hinges upon supporting and promoting the unglamorous but crucial work of ongoing repair, maintenance, and modernization of critical infrastructure systems. Events such as the 2003 blackout and the deadly 2007 collapse of the I-35 Mississippi River bridge in Minnesota illustrate that U.S. infrastructure systems are aging and, as they do, becoming more fragile and prone to failures. In each case, it was a lack of repair and maintenance that was the cause of failure, not intentional attack (Nye, 2010: 180; Patterson, 2010). But repair and maintenance are the key to resilient systems, not only because they reduce the fragility of those systems and thus help to prevent failures in the first place, but also because they promote learning and adaptation among the human repair crews that will be the first responders when failures do occur (Graham and Thrift, 2007: 5, 14; Nye, 2010: 189). Thus, instead of "think[ing] of the grid as a fortress to be protected at every point" (Nye, 2010: 197) by a central authority against total collapse caused by a hypothetical cyberattack, we should invest in the more mundane, ongoing, and essentially decentralized work of repair and maintenance that are the true source of resilient infrastructures (Graham and Thrift, 2007: 9–10). This warning against a fortress mentality and centralization of authority should apply to cyberspace itself and call into question proposals to "re-engineer the Internet" largely under the direction of the military (McConnell, 2010).

Disaster researchers have shown that victims are often themselves the first responders and that centralized, hierarchical, and bureaucratic responses can hamper their ability to respond in the decentralized and self-organized manner that has often proved to be more effective (Quarantelli, 2008: 895–6). In the case of preventing or defending against cyberattacks on critical infrastructure, we must recognize that most cyber and physical infrastructures are owned by private actors. Thus, a centralized, military-led effort to protect the fortress at every point will not work. A combination of incentives, regulations, and public-private partnerships will be necessary. The owners and operators of our critical infrastructures are on the front lines and will be the first responders. They must be empowered to act. Similarly, if the worst should occur, average citizens must be empowered to act in a decentralized, self-organized way to help themselves and others. In the case of critical infrastructures like the electrical grid, this could include the promotion of alternative energy generation and distribution methods. In this way, "Instead of being passive consumers, [citizens] can become actors in the energy network. Instead of waiting for blackouts, they can organize alternatives and become less vulnerable to either terror or natural catastrophe" (Nye, 2010: 203).

As noted earlier, cybersecurity and conflict have turned out to be much more than a technical challenge. Addressing the technical aspects of cybersecurity, as well as improving our ability to mitigate the effects of a large-scale attack should it occur, will require addressing the broader social, political, and cultural aspects of the problem. We must work to build strong communities, economies, and good governance locally and nationally. Just as more resilient technological systems can better respond in the event of failure, so too are strong social systems better able to respond in the event of disaster of any type. Historians and disaster researchers alike have documented that the response of individuals and groups in disaster situations largely depends on larger structural conditions in existence before the disaster itself. Communities that have weaker social ties among members, have corrupt or ineffective government and law enforcement, and suffer from economic hardship prior to a disaster will find it more difficult if not impossible to respond effectively in a time of crisis (Nye, 2010: 185; Lakoff, 2006; Alexander, 2006).

As the *Bulletin of Atomic Scientists* noted in January 2019, the erosion of trust in government, experts, and our fellow citizens fostered by malicious online influence campaigns, whether foreign or domestic in origin, makes it increasingly difficult to address other serious threats like nuclear warfare and climate change. We can add the technical aspects of cybersecurity threats to that list. For example, a number of recent reports note the lack of progress on election cybersecurity in a climate of U.S. domestic political chaos (Fazzini, 2019; Marks, 2019; Carney, 2019). What's more, a recent U.S. Senate report documented the abysmal state of cybersecurity at many U.S. federal agencies, including the Department of Homeland Security, perhaps the most important agency outside the military that should be leading the way (Cimpanu, 2019). It is hard to see how any of this gets fixed in a climate of political rancor fueled in part by malicious, cyber-enabled information operations.

In the end, improving the way we think, speak, and act with respect to cybersecurity will require moving beyond the kind of "scared straight" mentality found in cyber-doom rhetoric. Not only have cyber-doom scenarios turned out to be poor descriptors and predictors of real cyber threats, but also there is valid concern that they may have led us astray. Scenarios and metaphors will always play a role in helping us to think through the potential impacts of new technologies for societies, economies, and warfare. Nonetheless, cyber-doom rhetoric too often works to divert our attention away from reality and towards expectations of hypothetical, worst-case futures. In doing so, cyber-doom rhetoric makes it "difficult to distinguish fact from fiction," sometimes fooling us into believing that "it doesn't matter and they're the same after all" (Schwartau, 1991). This book has argued that the differences between fact and fiction do matter and so does the language and rhetoric we use to sort out those differences. In the wake of 2016, it is clear that we must do better in this regard. Working to move beyond the use of cyber-doom rhetoric will be critical to doing so.

References

Alexander, D. (2006) *Symbolic and Practical Interpretations of the Hurricane Katrina Disaster in New Orleans*. Presentation at Understanding Katrina: Perspectives from the Social Sciences, the Forum of the Social Science Research Council. New York: Social Science Research Council. Online. Available: <https://understandingkatrina.ssrc.org/Alexander/>.

Axelrod, R. (2014) "A Repertory of Cyber Analogies," in Goldman, E.O. and Arquilla, J. (eds) *Cyber Analogies*, Monterey, CA: Naval Postgraduate School, pp. 108–16. doi:10.21236/ada601645.

Banach, S.J. and Ryan, A. (2009) "The Art of Design: A Design Methodology," *Military Review*, 89, 2: 105–15.

Barnes, J. and Gibbons-Neff, T. (2019) "U.S. Carried Out Cyberattacks on Iran," *New York Times*, 22 June 2019. Online. Available: <www.nytimes.com/2019/06/22/us/politics/us-iran-cyber-attacks.html> (accessed 22 June 2019).

Boyd, J.R. (1987a) "A Discourse on Winning and Losing," unpublished manuscript.

———. (1987b) "Strategic Game of ? And ?," unpublished manuscript.

Brantly, A.F. (2016) *The Decision to Attack: Military and Intelligence Cyber Decision-Making*, Athens, GA: University of Georgia Press.

Brunner, E. and Dunn Cavelty, M. (2009) "The Formation of In-Formation By the Us Military: Articulation and Enactment of Informatic Threat Imaginaries on the Immaterial

Battlefield of Perception," *Cambridge Review of International Affairs*, 22, 4: 629–46. doi:10.1080/09557570903325454.

Cardon, E.C. and Leonard, S. (2010) "Unleashing Design: Planning and the Art of Battle Command," *Military Review*, 90, 2: 2–12.

Carey, J.W. (1989) *Communication as Culture: Essays on Media and Society*, New York: Routledge. doi:10.4324/9780203928912.

Carney, R. (2019) "Senate GOP Blocks Election Security Bill," *The Hill*, 25 June 2019. Online. Available: <https://thehill.com/blogs/floor-action/senate/450334-senate-gop-blocks-election-security-bill> (accessed 25 June 2019).

Charney, S. (2010) *Collective Defense: Applying Public Health Models to the Internet*, Redmond, WA: Microsoft Corp.

Cimpanu, C. (2019) "Report Shows Failures at Eight us Agencies in Following Cyber-security Protocols," *ZDnet*, 26 June 2019. Online. Available: <www.zdnet.com/article/report-shows-failures-at-eight-us-agencies-in-following-cyber-security-protocols/> (accessed 26 June 2019).

Department of Homeland Security. (2011) *Enabling Distributed Security in Cyberspace: Building a Healthy and Resilient Cyber Ecosystem With Automated Collective Action*, Washington, DC: Department of Homeland Security.

Dunn Cavelty, M. (2008) *Cyber-Security and Threat Politics: U.S. Efforts to Secure the Information Age*, New York: Routledge. doi:10.4324/9780203937419.

Dynes, R. (2006) "Panic and the Vision of Collective Incompetence," *Natural Hazards Observer*, 31, 2.

Elkus, A. (2010) "Complexity, Design, and Modern Operational Art: Us Evolution or False Start?" *Canadian Army Journal*, 13, 3: 55–67.

Fazzini, K. (2019) "Forget Mueller: Our Pants Are Still Down on Election Security, and Facebook Can't Save Us," *CNBC.com*, 29 May 2019. Online. Available: <www.cnbc.com/2019/05/29/mueller-cybersecurity-and-election-risk-facebook-cant-save-us.html> (accessed 29 May 2019).

Finn, J. (2016) "Cyber Security Expert Puts Growing Crisis Into Perspective," *College of Wooster Press Release*, 15 February 2016. Online. Available: <www.wooster.edu/news/releases/2016/february/recap-singer/index.php> (accessed 15 February 2016).

Goldman, E.O. and Arquilla, J. (eds) (2014) *Cyber Analogies*, Monterey, CA: Naval Post-graduate School. doi:10.21236/ada601645.

Graham, S. and Thrift, N. (2007) "Out of Order: Understanding Repair and Maintenance," *Theory, Culture & Society*, 24, 3: 1–25. doi:10.1177/0263276407075954.

Haraway, D.J. (1991) *Simians, Cyborgs, and Women: The Reinvention of Nature*, New York: Routledge. doi:10.4324/9780203873106.

JASON. (2010) *Science of Cyber-Security*, McLean, VA: JASON, The MITRE Corporation.

Kostyuk, N. and Zhukov, Y.M. (2019) "Invisible Digital Front: Can Cyber Attacks Shape Battlefield Events," *Journal of Conflict Resolution*, 63, 2: 317–47. doi:10.1177/0022002717737138.

Lakoff, A. (2006) *From Disaster to Catastrophe: The Limits of Preparedness*. Presentation at Understanding Katrina: Perspectives from the Social Sciences, the Forum of the Social Science Research Council. Online. Available: <https://understandingkatrina.ssrc.org/Lakoff/>.

Lapointe, A. (2011) *When Good Metaphors Go Bad: The Metaphoric "Branding" of Cyberspace*, Center for Strategic and International Studies.

Lawson, S. (2010) "These Aren't the Objective Cybersecurity Commission Members You're Looking For," *Forbes.com*, 20 April 2010. Online. Available: <www.forbes.com/sites/firewall/2010/04/20/not-the-cybersecurity-commission-you-are-looking-for/> (accessed 20 April 2010).

———. (2012) "Putting the 'War' in Cyberwar: Metaphor, Analogy, and Cybersecurity Discourse in the United States," *First Monday*, 17, 7. doi:10.5210/fm.v17i7.3848.

———. (2013) "The US Military's Social Media Civil War: Technology as Antagonism in Discourses of Information-Age Conflict," *Cambridge Review of International Affairs*, 27, 2: 226–45. doi:10.1080/09557571.2012.734787.

———. (2014) *Nonlinear Science and Warfare: Chaos, Complexity, and the U.S. Military in the Information Age*, London: Routledge. doi:10.4324/9780203766446.

Lewis, J.A. (2006) "The War on Hype," *San Francisco Chronicle*, 19 February 2006. Online. Available: <https://articles.sfgate.com/2006-02-19/opinion/17283144_1_cyber-attack-pandemic-avian-flu> (accessed 19 February 2006).

———. (2010) "The Cyber War Has Not Begun," unpublished manuscript.

Libicki, M.C. (1997) *Defending Cyberspace, and Other Metaphors*, Washington, DC: U.S. Government Printing Office.

Liles, S. (2010) "Cyber Warfare: As a Form of Low-Intensity Conflict and Insurgency," in Czosseck, C. and Podins, K. (eds) *Conference on Cyber Conflict Proceedings 2010*, Tallinn, Estonia: CCD COE Publications, pp. 47–58.

Lindsay, J.R. (2013) "Stuxnet and the Limits of Cyber Warfare," *Security Studies*, 22, 3: 365–404. doi:10.1080/09636412.2013.816122.

———. (2015) "The Impact of China on Cybersecurity: Fiction and Friction," *International Security*, 39, 3: 7–47. doi:10.1162/isec_a_00189.

MacKinnon, D. (2019) "Doug MacKinnon: Will You Survive the Coming Blackout?," *FoxNews.com*, 23 June 2019. Online. Available: <www.foxnews.com/opinion/doug-mackinnon-survive-coming-blackout> (accessed 23 June 2019).

Marcus, D. (2019) "Why Trump's Cyber Attack on Iran Was the Right Move," *The Federalist*, 23 June 2019. Online. Available: <https://thefederalist.com/2019/06/23/trumps-cyber-attack-iran-right-move/> (accessed 23 June 2019).

Marks, J. (2019) "The Cybersecurity 202: Even a Voting Machine Company Is Pushing for Election Security Legislation," *The Washington Post*, 10 June 2019. Online. Available: <www.washingtonpost.com/news/powerpost/paloma/the-cybersecurity-202/2019/06/10/the-cybersecurity-202-even-a-voting-machine-company-is-pushing-for-election-security-legislation/5cfd75691ad2e5122b87c567/?noredirect=on&utm_term=.574ad024fd19> (accessed 10 June 2019).

McConnell, M. (2010) "Mike McConnell on How to Win the Cyber-war We're Losing," *The Washington Post*, 28 February 2010, B01.

Mongoven, A. (2006) "The War on Disease and the War on Terror: A Dangerous Metaphorical Nexus?" *Cambridge Quarterly of Healthcare Ethics*, 15, 4: 403–16. doi:10.1017/s0963180106060518.

Nakashima, E. (2019) "Trump Approved Cyber-strikes Against Iran's Missile Systems," *The Washington Post*, 22 June 2019. Online. Available: <www.washingtonpost.com/world/national-security/with-trumps-approval-pentagon-launched-cyber-strikes-against-iran/2019/06/22/250d3740-950d-11e9-b570 6416efdc0803_story.html?noredirect=on&utm_term=.5cfc1e4b62ed> (accessed 22 June 2019).

Nye, D.E. (2010) *When the Lights Went Out: A History of Blackouts in America*, Cambridge, MA: MIT Press. doi:10.7551/mitpress/8252.001.0001.

Patterson, T. (2010) "U.S. Electricity Blackouts Skyrocketing," *CNN.com*, 15 October 2010. Online. Available: <www.cnn.com/2010/TECH/innovation/08/09/smart.grid/index.html?hpt=Sbin> (accessed 15 October 2010).

Quarantelli, E.L. (2008) "Conventional Beliefs and Counterintuitive Realities," *Social Research: An International Quarterly*, 75, 3: 873–904.

Schmitt, C.J.F. (2006) *A Systemic Concept for Operational Design*, Quantico, VA: Marine Corps Combat Development Command.

Schwartau, W. (1991) *Terminal Compromise*, Seminole, FL: Inter-Pact Press.

Simpson, C. (1994) *Science of Coercion: Communication Research and Psychological Warfare, 1945–1960*, New York: Oxford University Press.

USSTRATCOM. (2009) *The Cyber Warfare Lexicon: A Language to Support the Development, Testing, Planning, and Employment of Cyber Weapons and Other Modern Warfare Capabilities*, Offutt Air Force Base, Nebraska: USSTRATCOM.

Valeriano, B., Jensen, B.M. and Maness, R.C. (2018) *Cyber Strategy: The Evolving Character of Power and Coercion*, New York: Oxford University Press. doi:10.1093/oso/9780190618094.001.0001.

Valeriano, B. and Maness, R.C. (2015) *Cyber War Versus Cyber Realities: Cyber Conflict in the International System*, London: Oxford University Press. doi:10.1093/acprof:oso/9780190204792.001.0001.

von Ghyczy, T. (2003) "The Fruitful Flaws of Strategy Metaphors," *Harvard Business Review*, September.

Walton, D.N. (2000) *Scare Tactics: Arguments That Appeal to Fear and Threats*, Boston: Kluwer Academic Publishers. doi:10.1007/978-94-017-2940-6.

Wass de Czege, B.G.H. (2009) "Systemic Operational Design: Learning and Adapting in Complex Missions," *Military Review*, January–February: 2–12.

Index

Note: Page numbers in **bold** indicate a table.

@cybersquirrel1 77

9/11 16, 63; al-Qa'ida- Iraq connection and 52; cyber 5, 106; analogies to 165; comparisons to 57; vs. cyber-kinetic attacks 16; economic impact of 106; NCW theorists on 90–1; wargames and 43
1995 Oklahoma City bombing 9, 48

Adams, John 78
Adas, Michael 79
aerial bombardment 84, 86, 102, 103, 106, 144
Afghanistan 9, 172
airpower theorists 86, 103
Alexander, Keith 11, 13–14, 57
al-Qa'ida 91
al-Qa'ida- Iraq connection, and 9/11 52
Altheide, David 40
amateur radio broadcast/operators 87
America Can Win (Hart & Lind) 89
American Blackout (National Geographic Channel) 47, 77, 130–31, 133
American democracy, role of communication technology on 84
Andres, Richard 47
anonymity network 1
anti-virus usage 130
appropriation 56; of fear 49–52; and historical events 41
Arab Spring conflicts 143
Aristotle, and fear appeal messages 128–29
Armageddon, aerial 87, 104
Army and Marine Corps 89
Army's Foreign Military Studies Office 165

Arquilla, John 15, 45, 165, 166, 168, 174; "Cyberwar is Coming!" 61, 165
articulations 19–20
Assange, Julian 4
Atkin, Charles 52, 53, 131
audience resistance 53
authoritarian regimes/states 141
autonomous technology 78

Barnett, T.P.M. 92
Barzaska, I. 115
Beck, Ulrich 85
Bell, Daniel 81
Beniger, James 81
Beyerchen, Alan 135
Bipartisan Policy Center 46
Black, Cofer 63
Black Hat hacker conference 63
"black" operations 174–179
Blank, Stephen 108
Blunden, Bill 37, 109
bombardment; aerial 84, 86, 102, 103, 106, 144; strategic 103–4
Borg, S. 111
Boyd, John 89
Broderick, Matthew 48
Bulletin of the Atomic Scientists 6, 7
Bumgarner, J, 111
Bush, George H. W. 132
Bush, George W. 9

Caldwell, French 60
Carey, James 78, 79, 81
Carper, Tom 64
Carson, André 164
Carter, Ash 57, 65, 167

Castells, Manuel 81
Cavelty, Brunner 169
Cavelty, Myrian Dunn 7–8, 17–18, 169
Cebrowski, Arthur 90, 92
Center for Risk and Economic Analysis of
   Terrorism Events 106
Center for Strategic and International
   Studies 37
Charney; Scott 10
Chechen insurgency 172
China 51, 55, 58; cyber espionage/spies 5,
   142; and cyber threats 50; and economic
   espionage 15
Christian Science Monitor 43–4
Cilluffo, Frank 54
civilian populations 87
clandestine activities, program of 176–7
Clapper, James 6, 16, 18, 138
Clarke, I. F. 93
Clarke, Lee 106, 116
Clarke, Richard 16, 63, 148, 166; *Cyber
   War* 15
Clausewitzian 55, 88
Clinton, Bill 48
Clinton, Hillary 12
Clodfelter, M. 102
CNN 15, 16; *Cyber Shockwave* 47
Coats, Dan 51
COINdinistas/COIN doctrine 91, 92
Cold War 85, 90, 92, 93, 174, 178–9;
   metaphor and analogy of 50; nuclear
   standoff 104; political warfare 179;
   worst-case entrepreneurs 52
Collins, Susan 5, 64
Comey, James 164
communication, and fear appeals 65, 131
communication revolution 81
Computer Fraud and Abuse Act 48
computer hacking techniques 165
computerized systems analysis 93
computer revolution 81
conflagation activities 55–6
Congressional hearings 36, 38; on
   Russia 164
Congressional Research Service 106, 113
control revolution 81
Cooperative Cyber Defence Centre of
   Excellence 109
counterinsurgency (COIN) approach, to
   Iraq 91
covert operations 176
CrashOverride 117, 118, 119
credit card information 11

critical infrastructures/systems 77; chances
   of attacks on 54; history of failure of
   101–7; protection of 7; terrorist groups
   against 9; threats to 7, 13–14, 48, 59
CSIS Commission on Cyber Security for
   the 44th Presidency 10
CSO Online 130
"cuisinart effect," 41
cultural anxieties, and cyber-doom
   rhetoric 82
cyber 9/11 5, 15, 106; and cyber threats
   comparisons 51, 59; and Estonia 108;
   likelihood of 6
cyber Armageddon scenario 6
cyber arms race 43
cyberattacks; on Estonia 15; false reports
   of 59; types of 6
CyberCity scenarios 44
cyber conflict language, dangers of 136–7
cyber conflicts, exaggerations of 161
cyber crime, economic cost of 58
cyber-doom fear appeals 61–2
Cyber-doom rhetoric, and fear appeal
cyber-doom rhetoric 77–8, 94; and cultural
   anxieties 82, 127; and fear appeals
   65–6, 82, 132–3, 150; and Georgia
   cyberattacks 111; negative implications
   of 121; and the NotPetya malware attack
   118–19; and persuasion 65; reason for
   77; selling of fear of and 40; and U.S.
   cybersecurity discussions 185; use of
   metaphors and analogies in 100
cyber-doom scenarios 2–3, 6; *see also*
   cyber-doom rhetoric; media, mass;
   cultural trends and 66, 94–5; dangerous
   obsession with 161; fictional 5; and
   information and communication
   technologies (ICTs) 82; real or hype
   15–18, 56; rhetoric of 15–18; selling
   fear and 36–7, 39, 45–6, 52, 59–60, 127;
   software use and 60; use of metaphors
   and analogies in 141–145
cyber espionage 57; and U.S. economic
   security 57
Cybergeddon web-series 46
Cyber Guard exercise 61
cyber hacking operations 164
cyber incidents, exaggerations of 56
cyber Katrina 51
cyber-kinetic attacks 1, 16
cyber Pearl Harbor scenario 3, 5, 16, 55;
   appropriation of 49, 57, 59; and cyber
   threats comparison 51; fictional report

and 45; likelihood of 2, 7; and Panetta 44; selling fear and 52, 61
cybersecurity; concerns about 2; evolution of 8–15
Cyber Security and American Cyber Competitiveness Act (2011) 13
cybersecurity debate 17
Cybersecurity Information Sharing Act (2015) 5
cybersecurity policies and strategy; militarized 18; Obama administration and 11
Cyber Security Public Awareness Act (2011) 12, 57
cybersecurity threats; call for attention to 50–1; and economic impact 12; as a military concern 14; public- private partnership and 43; reasons for concern 13–14; and the USA 11; use of natural and man-made disasters and 51
Cyber Shockwave (CNN) 15, 46, 47
cyberspace 48, 54
Cyberspace Policy Review, Obama administration and 11
Cyberspace Solarium Commission 140
Cyber Squirrel 77
Cyber Statecraft Initiative 63
cyberterror/cyberterrorism 1–2, 17; *Cybergeddon* web series and 46; and Internet use 55
cyber threats 6; beliefs about 46; growth of 53–4
cyber tools 6–7
"Cyberwar is Coming!" (Arquilla & Ronfeldt) 61, 165
cyber war theorists 86
cyber war/warfare 54–5, 116, 162, 165–68; emotions and 37; in Estonia 107–10; in Georgia 110–13; in Iran 113–16; in Ukraine 116–20
cyber "weapons" 140–41

Danger on the Internet Highway (Fox News) 45–6
data breaches 54, 56; use of exaggerations and 57
"Day After Tomorrow" study 44
"Day After Tomorrow" wargame 59
Defense Advanced Research Projects Agency (DARPA) 147
Defense Cyber Strategy 55
Defense Information Systems Agency (DISA) 43

Defense Intelligence Agency 45
Democratic National Committee 44
denial of service (DDOS) attacks 110
denial of service attacks 55, 58, 163
Department of Defense Cyber Strategy 168
Department of Defense Strategy for Operating in Cyberspace 13, 14
Department of Defense Strategy for Operations in the Information Environment 170
Department of Homeland Security 44, 59
Derrida, Jacques 19
de Saussure, Ferdinand 19
deterministic thought/determinism 80–1, 141–3; soft 80, 82; technological 90; and the military 85
Deutch, John 50
Devost, M. G. 119
Die Hard franchise 47
digital computer, general purpose 88, 93
digital Pearl Harbor 57
digital sabotage 55
disasters, reactions to 104–5
DOD 167
DoD Cyber Strategy (2015) 15
doxing 55
Dragos 117, 118, 119
Dukakis, Michael 132
Dunn Cavelty, Myrian 7, 8, 17–18

economic espionage, and China 15
election interference, U.S. 164–5, 168
electrical blackouts 51–2, 59–60, 77, 105–6; appropriation of 52; in New York City 103, 105; war-time 102
electrical grids; aerial bombardment of 84; attacks on 116–17; effect of cyberattack on U.S. 47; technological pessimism 83
electricity/electrification 83; Victorian Age response to 82, 83
electronic communication/networks; early use of 84; technological pessimism and 83
electronic Exxon Valdez 51
electronic Pearl Harbor 7, 36, 37; *see also* cyber Pearl Harbor scenario
Eligible Receiver wargame 43, 50
Elmer, 61
Enhancing the Resilience of the Nation's Electricity System 119
Ergma, Ene 15, 16
Estonia: cyberattacks on 9, 15–16, 57, 107–10, 166

EternalBlue 117
European Recovery Program 175
Evera, Van 52
eWMDs 15
Exxon Valdez, electronic 51

fake news and disinformation, filtering
    of 147
"fantasy documents" 42
fear; as an agent of change 38; culture and
    politics 40–1; selling of 37
fear appeals 38–40, 52–3; *see also* fictional
    cyber-doom scenarios; appropriation of
    49–52; as communication 65, 127–8,
    131; and cyber-doom rhetoric 65–66;
    failure of 128–33; messages 128–9;
    research 129–31; unethical 131–3; use
    of exaggerations and 56–8
FedCyber conference 45
Federalist 78
fiction; military science 93; use of 64–5, 93
fictional cyber-doom scenarios 5, 45;
    impact on policymaking 47
Fifth Domain 3
"Fighting Terminal Terrorism,"
    (Schwartan) 36
First World War *see* WWI
Fitzgerald, Cantor 43
Flame attacks 64, 143
Fogarty, Stephen 145
Foreign Affairs (Lynn) 14
Foreign Policy magazine 51, 77; *Fox
    News*: *Danger on the Internet Highway*
    45–46, 47
Frank, Howard 59, 61
Freedman, Lawrence 94, 103
Fukushima nuclear disaster 51
Furedi, Frank 40, 42, 85
Future Shock (Toffler) 81
"Future Warfare. . ." (Mattis & Hoffman)
    172

Galeotti, Mark 165
gaming and simulation 93
Georgia, cyberattacks on 9, 58, 110–13
Germany, U.S. bombardment of 102–3
Gibson, William, *Neuromancer* (Gibson) 48
Giddens, Anthony 85
Giles, Keir 112
Gingrich, Newt 44
Glassner, Barry 40, 62, 142
global civilization, cyberattacks against 16
globalization 90

Google case, and Chinese government
    cyberattack allegations 12
Gorelick, Jamie 50
Gourley, Bob 45, 60, 61
Government Accountability Office (GAO)
    64, 173
government secrets, theft of 10, 11, 12–13
GovTech 52
Greek democratic republic 78
Greenberg, Andy 117, 118–19, 120
Gross, Michael 64
guerrilla warfare 172

Hacker, Bart 87
hackers 1, 108
"Hackers Take Down the Most Wired
    Country in Europe" 108
hacktivism 55
Hart, B.H. Liddell 87; *Paris, or the Future
    of War* 86
Hart, Gary, *America Can Win* 89
Hasian, Marouf 80, 141
Hayden, Michael 46, 55, 63, 118
Healey, Jason 63, 109, 149, 150
Hertling, Mark 164
Hiroshima 102, 114
Hoffman, Frank, "Future Warfare." 172
Homeland Security Policy Institute 54
Hoover Institution 15
Horton, Willie 132
hostile intelligence threat 8, 9
Hughes, Thomas 78, 79–80
human history, periodization of 90
Hunker, Jeffrey 51
Hurricane Katrina 106–7, 137
Hussain, Junaid 149
hybrid warfare 143, 162, 172–4; Russia
    and the U.S.A. 164; term usage 172

ICT-enabled weapon systems 89
imaginaries 19–20
"Inauguration of Organized Political
    Warfare, The" (Kennan) 174
India 60
Indian Ocean tsunami 16, 51
"industrial fabric" theory 86
industrial fabric theory 86
industrialization/industrialized societies 86
industrial-mechanized warfare 86
inevitable future, the 61
information, integrity of /theft of 6
Information Age 19, 88, 92
Information-Age security imaginary 18–19

Information-Age warfare 44, 86, 88, 90, 92
Information-Age weapons 88
Information-Age worldview 81
informational damage, integrity of /theft of 10, 14–15
information and communication technologies (ICTs) 81, 82; military ambivalence and 85, 91
information assurance 87
information revolution 81
information warfare 19, 162–3, 166; term usage 168–72
Information Warfare Operations Command 145
intellectual property, theft of 10, 11, 12–13, 57
international relations, cyber powers and 100
International Strategy for Cyberspace, Obama administration and 12
Internet freedom 12, 18
Internet use, as cyberterrorism 55
Iran 9; cyberattacks on 64, 113–16
Iranian enrichment program 114–15
Iraq 89, 90, 172; 2nd U.S. invasion of 166; WWD and justification for war 92
Islamic State (ISIS) 91; cyberattacks against 65
ISIS terrorist attacks (Paris) 51

Jaitner, M. 163
Jefferson, Thomas 78–9, 84
Johnson, Lyndon, "War on Poverty," 79
Johnson, M. 134
Journal of Commerce 43

Kaplan, Fred 48, 59, 62–3, 108, 148; and Georgia cyberattacks 111; and Stuxnet 114; Stuxnet 143
Kaspersky Labs 117
Kastenberg, J.E. 111
Katrina, cyber 51, 107
Kelly, Kevin 81
Kennan, George F. 175, 178; "The Inauguration of Organized Political Warfare" 174
Khobar Towers bombing (Saudi Arabia) 50
Kilcullen, David 91, 92
"kinetic" cyberattack 1
Klaas, Brian 162, 163, 173
Knake, Robert 16, 116, 166; *Cyber War* 15
Konvitz, Joseph 102–103
Korns, S.W. 111
Kurtz, Paul 51
Kyl, Jon 12, 57

Lakoff, G. 134
Langevin, James 10
language, bad (cyber) 136–37
Lapointe, A. 139, 140
Lasker, Larry 49
Lee, Robert 7, 118, 119
LeMay, Curtis 88
Lewis, James 6, 37, 54, 105, 108, 137–138, 138; CSIS Commission on Cyber Security and 10; cyberwar/ warfare and 199
Libicki, Martin 119, 139, 140
Lieberman, Joseph 64, 147
Lin, Herb 7
Lind, William, *America Can Win* 89
Lindsay, Jon 114
Lunday, Kevin 61
Lynn, William 13, 54; *Foreign Affairs* 14

Maersk 117–18
Maness, R. C. 109, 110, 112, 115, 163
maneuver warfare 89, 93
Mansfield-Devine, S. 109
Marine Corp's Amphibious Warfare School 89
Markoff, John 3
Marsh, Robert 50
Marshall Plan, the 175, 183
Marx, Leo 79, 80, 82, 85
mass communication 84–5
mass media fear appeals 47
Mattis, James, "Future Warfare" 172
McBurney, Peter 139
McCabe, M. 119
McCain, John 164
McCaul, Michael 10
McConnell, Mike 16, 49, 51, 63
McKew, Molly 164
McMaster, H.R. 163, 172, 173
mechanization 86, 87
mechanized warfare theorists 86
media, mass, selling fear and 46–7, 60
Merck 118
metaphorical idolatry 139–41
metaphors, use of 133–6
metaphors of war 145
militarized cybersecurity policies 18
military computerization 88
military cyberspace policy 168
military-entertainment complex 94
military history, gaming and simulation and 93
military isomorphism 19

Military Reform Caucus 89
military science fiction 93
military-technical revolution 89–90, 92
military thinkers, and technology 85–94
Military Times 44, 61
momentum, high 80
Mongoven, Ann 137
moral panic 85
movies, about cybersecurity 47
Mueller, Robert 46

Nakasone, Paul 145
Napolitano, Janet 16, 51
NASDAQ 77
National Cyber Research Park 43
National Cyber Security Division
   (Homeland Security) 15, 108
*National Geographic Channel, American
   Blackout* 47, 77, 130–31, 133
national news organizations 84
National Policy on Telecommunication 48
National Security Agency (NSA) 4, 16
National Security Council's Office of
   Transnational Threats 51
NATO Co-operative Cyber Defence Centre
   of Excellence 1, 16, 108, 112
natural and man-made disasters;
   comparisons and 16; mass panic and
   128; as metaphor and analogy of 51
Naval War College games 43
"necessitous" rhetorics/thought 80, 141
netwar 166, 174
"network-centric warfare" (NCW) 90–1
Neuromancer (Gibson) 48
new economy 81
news media, role in developing culture of
   fear 42
New York Times 3, 114, 140, 150
"Nightmare Round" 45
non-cyber events, appropriation of 49–52
nonfiction, instrument of fear 42
Noon, David 139, 142
North American Aerospace Defense
   Command (NORAD) 48, 64
North Korea 5, 55
Norton Antivirus 46
NotPetya malware attack 116, 117,
   117–20; clean-up of 117–18;
   exaggerations of 118
NSDD-145 48
nuclear weapons 58; analogies 57; and
   cyber threats comparison 50
Nunn, Sam 44
Nye, Joseph S. 7, 79, 80, 102, 106

Obama, Barack 5, 9–10, 51; speech on
   cybersecuirty 11–12; cyberattacks
   response and 54
Obama administration and; cybersecurity
   policy and strategy 11, 13; *Cyberspace
   Policy Review* 11; *International Strategy
   for Cyberspace* 12
Office of Net Assessment (ONA) 89, 168
Office of Personnel Management (OPM),
   hacking of 5, 51, 55
official nonfiction 42
Offset Strategy 89
Oklahoma City bombing 9, 48
Olympic Games 63
Opel, A. 22, 61
Operation Desert Storm 89
Organization for Economic Cooperation
   and Development (OECD) report 17
Ottis, R. 109

Panetta, Leon 2, 3, 5, 7, 16; criticism of his
   speech 137–8; and cyber Pearl Harbor
   scenario 44, 46, 51; shock therapy
   analogy and 38; and Stuxnet 114
Paris, or the Future of War (Hart) 86
Parkes, Walter 49
PCCIP report 61
Pearl Harbor, analogies to 57, 165;
   appropriation and 49, 57
perception management 19
periodization, human development and 81
pessimism, technological 80, 83, 90
Petya ransomware 117
Pfau, Michael 39, 128
physical warfare, and cyberattacks 18
political warfare 174, 175–8
Pollard, N.A. 119
possibilistic thought 144
post-industrial/network society 81
power grids, attacks on 52, 58–9
President's Commission on Critical
   Infrastructure Protection (PCCIP) 9,
   48, 50
probability neglect, war metaphors and 141–5
product placement, and cyber terrorism
   rhetoric 46
projection; U.S. cybersecurity debate and
   62–5; use of exaggerations and 62–5, 92
propaganda operations 6, 14; and
   cyberattacks 17
psychological warfare 163, 174
public-private Cyber Guard exercise 44
public- private partnership; in
   cybersecurity 43, 44

radio broadcast/operators, amateur 87
Raduege, Harry 10, 149
RAND Corporation 44, 88, 93
Ranum, Marcus 37
Reagan, Ronald, reaction to *WarGames* 48
Reagan, T. 119
Reagan build-up 94
Real Clear Politics (Walworth) 140
Reid, Harry 13
revolution; computer 81; control 81;
    information 81; military- technical
    89–90, 92
rhetorical software 60–1
Rid, Thomas 17, 114
"risk society" thesis 85
Robin, Corey 39, 41
Rochlin, Gene 81
Rogers, Michael 5, 57, 58, 61, 118, 144,
    164; term usage and 174
Rona, Thomas 168
Ronfeldt, David 166, 168, 174; "Cyberwar
    is Coming!" 165
Rothkopf, David 51
Royall, Kenneth 175
Rumsfeld, Donald 92
Russia 51, 55, 116; and the U.S.A., hybrid
    warfare and 173
Russia-Georgia war 110–113
Russian cyberattackers 59
Russian cyber-interference/ operations 6,
    14–15, 50, 65; political nature of 17, 44,
    161, 162, 164–5; and U.S. presidential
    elections 142, 145, 179
Russian hackers 9
Russian Internet Research Agency 145
Russia-Ukraine conflict 116–20

Sacramento Bee 65
Sanger, David 140
Saudi Aramco, cyberattack on 64
Schneier, Bruce 16, 108
Schön, Donald 134
Schulte, Stephanie 38, 47
Schwartau, Winn 3, 7, 44, 45, 56;
    cyberterrorism warnings of 56;
    electronic Pearl Harbor 47, 49;
    "Fighting Terminal Terrorism," 36;
    Terminal Compromise 36
*Securing Cyberspace for the 44th
    Presidency* 10
security imaginaries 19–20, 161
selective strike 173
September 11, 2011 *see* 9/11
Shaheen, Jeanne 164

Sherman, William C. 86, 87
shock therapy 3
Simon, Linda 83
Simpson, Christopher 177, 178
Skyfall 2, 47; and real-life cybersecurity
    issues 4
slum districts 87
Smeets, Max 150
Sneakers (film) 48
Snopes.com 60
Snowden, Edward 4, 57
Snyder, Timothy 163, 174, 180
social imaginary 19–20
social media, hacking of 55
societies, resilient of 120–21
sociotechnical revolution 91
socio-technical systems; fear of
    technology and 82–5; impact of 79–80;
    technological pessimism and 84
soft determinism 82
Solar Sunrise 59
Sony Pictures Entertainment, hacking of
    5, 55, 57
Soviet Union 104
Speier, Jackie 164
Stefanik, Elise 164
Stewart, Vincent 146
Stohl, Michael 16
Stone, Adam 3
Strategic Air Command 88
strategic ambiguity 52–3
strategic bombardment 103–4
Stuxnet 63, 113–16, 143; attack 60, 148;
    used as example of cyber threat 64, 114
Subcommittee on Technology and
    Competitiveness 36
Sullivan, Gordon R. 90
Sunstein, Cass 40, 42
Super Storm Sandy 16, 51
Swiss Federal Institute of Technology 17
Symantec 46, 51
systems sciences 88, 93

technium 81
technological determinist mindset 78,
    90, 102
technological pessimism 80, 83, 90
technology; authoritarian states and 141;
    cultural and historical anxieties about
    78; fear appeal 94; fear of 85; military
    thinking and 85–93
technopanic/ technophobia 85
Telefon 47
Terminal Compromise (Schwarrtau) 36

terrorist attacks, and cyber threats
    comparison 51
Third Wave, The (Toffler) 81
Toffler, Alvin & Heidi 44, 90; *Future
    Shock* 81; *Third Wave, The* 81; *War and
    Anti-War* 81
Trump, Donald 163
Trump administration 149, 164

Ukraine 58, 61, 162; cyberattacks on
    116–20, 164
United Nations' International
    Telecommunications Union 51
U.S. Army's Strategic Studies Institute 108
U.S. Central Command website 149
U.S. Congress, cybersecurity threat
    concerns and 12–13
USCYBERCOM 144, 146–7, 167, 168
U.S. Cyber Command 11, 13, 14, 57
U.S. Cyber Consequences Unit 111
U.S. cybersecurity debate 61; projection
    and **62–5**; rhetorical 61
U.S. cybersecurity policy; debate about 37;
    review of 10
U.S. cyber war policy, using projection 62
U.S. Department of Transportation's
    (DOT) campaign 131
U.S. economic competitiveness 142
U.S. economic security 57
U.S. National Academy of Sciences
    report 119
U.S. National War College 47
U.S. Navy 87
U.S. PPD-20 document 116
U.S. presidential elections 2, 5; and Russian
    cyber operations 65, 142, 145, 161, 162,
    164, 164–5, 179; term usage for 168
U.S. public policy debate; and the fear
    factor 37; strategic ambiguity and 53
U.S. Strategic Bombing Survey 102
U.S. websites, denial of service attacks 55

Valeriano, B. 109, 110, 112, 115
Vanity Fair 114
Victorian Internet 84, 175
Vietnam experience 88–9

vulnerabilities; conflagration of 55–6; fear
    appeal and 55–6

Wall Street Journal 120
Walton, Douglas 38, 131, 131–2
Walworth, Andrew, *Real Clear
    Politics* 140
war; Clausewitzian definition of 55;
    conflagation of 55–6; metaphors of
War and Anti-War (Toffler) 81
wargames 46
WarGames 4, 38, 47–8, 49
war metaphors; as metaphorical idolatry
    139–41; and the probability of neglect
    141–5; as source of distraction 137–9;
    use of 137–9
"War on Poverty," 79
war on terror 139, 144
Washington Post 43, 45, 46, 60, 162
Watts, Clint 163
WDD, analogies 57, 58
Weaver, N. 119
websites, defacement of 110–11
Whitehouse, Sheldon 12–13, 57
Wiener, Norbert 81, 90
WikiLeaks 4, 55
Winner, Langdon 80
Wired 59, 81, 108, 117, 119
wire devils *see* hackers
Wittenstein, Edward 3
Work, Bob 167
World Trade Center bombing 50
World War II *see* WWII
worst-case entrepreneurs 41, 45, 56,
    65, 188; and fear appeal messages
    52, 56; government as 42; and use of
    conflagration tactics 56; use of metaphor
    and analogy of 56
WWD 92
WWI 79, 84, 93
WWII 79, 85, 87, 88, 93; analogies to 141,
    142; destruction and 102

Yoran, Amit 15, 16, 56, 108

Zetter, Kim 118